12/25
25

AI摄影与后期制作101例
（100集视频课）
罗巨浪　周冰渝 编著
清華大學出版社
北京

内 容 简 介

人工智能时代，AI技术的应用给各行各业带来巨大的变化。而Midjourney、Stable Diffusion、文心一格等AI绘画软件的出现，更是为艺术创作领域注入了新的活力，极大地拓展了创意表达的边界。

本书就以Midjourney为基本工具，通过101个AI摄影案例，详细介绍了AI技术在摄影领域的应用。全书分为基础篇、实践篇和后期制作篇3篇，其中基础篇简单介绍了几款AI辅助摄影软件及摄影的基本知识，旨在帮助读者掌握构图、光线运用等摄影关键知识。实践篇为全书的核心，用10章的篇幅介绍了Midjourney在自然风景、静物、人像、创意摄影、特效摄影等不同摄影类型中的摄影案例，让读者掌握使用AI工具进行摄影创作的方法和技巧。后期制作篇包含换脸和后期处理两章，介绍了AI换脸功能的使用和图像后期制作的方法，旨在帮助读者掌握如何通过后期处理，提升AI摄影作品的艺术表现力。

本书内容丰富，案例多，不但涵盖了AI摄影的基础操作，还涉及了AI摄影的主要应用场景，适合所有摄影爱好者、摄影师以及对AI绘画感兴趣的读者学习。

图书在版编目（CIP）数据

AI 摄影与后期制作 101 例：100 集视频课 / 罗巨浪，周冰渝编著. -- 北京：清华大学出版社，2024.11.

ISBN 978-7-302-67650-8

Ⅰ. TP391.413

中国国家版本馆 CIP 数据核字第 2024YG2217 号

责任编辑： 杜　杨

封面设计： 墨　白

责任校对： 徐俊伟

责任印制： 沈　露

出版发行： 清华大学出版社

网　　址： https://www.tup.com.cn，https://www.wqxuetang.com

地　　址： 北京清华大学学研大厦 A 座　　**邮　　编：** 100084

社 总 机： 010-83470000　　**邮　　购：** 010-62786544

投稿与读者服务： 010-62776969，c-service@tup.tsinghua.edu.cn

质 量 反 馈： 010-62772015，zhiliang@tup.tsinghua.edu.cn

印 装 者： 北京博海升彩色印刷有限公司

经　　销： 全国新华书店

开　　本： 170mm × 240mm　　**印　　张：** 13　　**彩　　插：** 2　　**字　　数：** 328 千字

版　　次： 2024 年 12 月第 1 版　　**印　　次：** 2024 年 12 月第 1 次印刷

定　　价： 79.80 元

产品编号：109745-01

前言 PREFACE

人工智能（Artificial Intelligence，AI）是当今时代新兴的颠覆性技术，虽然还在成长与摸索中，但仍然改变着人类的生产生活方式和思维模式，对经济发展、社会进步等方面也产生了影响。近几年来，多种多样的 AI 软件横空出世。通过简单的指令提示，就能让 AI 在短时间内创作出相应的图像作品。AI 技术的飞速发展同样引发了摄影行业的颠覆性革命。

摄影因 AI 变得简单、便捷。如今足不出户就可以借助 AI 拍到自己想尝试的风格，原本必须由人工操作、手动后期、效率低下的摄影模式将演变为智能化、自动化、高效率的制作方式。AI 摄影不仅适用于抖音、小红书等平台活跃的自媒体达人，也同样适用于普通人。AI 摄影将成为影响时代发展的必要推力。

为了满足广大摄影爱好者的学习和实际工作需求，作者创作了《AI 摄影与后期制作 101 例（100 集视频课）》。本书具有以下三大特色。

- 场景式学习：根据摄影画面的常见需求，本书设计了多种使用场景，如自然风景、场景、静物、动物、人像、创意摄影、不同艺术风格、景别控制、特效摄影、不同构图等。读者可以根据自己的需求，挑选不同场景下的案例进行学习，实现“所学即所用”。
- 案例式学习：本书共展示了 101 个案例。读者在学习案例的基础上，可以通过替换提示词的方式创作出属于自己的新作品。在反复的实操训练中，快速掌握 AI 摄影的技巧。
- 工作流式学习：为了帮助读者提高作品的完成度，在 AI 摄影的基础上，本书还结合了 Photoshop 等常用的辅助软件，讲解了图像的后期处理与应用。这种方法可以引导读者从输入提示词到后期加工，循序渐进地完成自己的摄影作品。

无论读者是初学者还是有经验的用户，本书都能成为你学习和应用 AI 摄影的得力工具，助力你的工作和生活。

尽管本书已经过作者和编辑的精心审读，但限于时间，难免有疏漏之处，望读者体谅包涵，不吝赐教。

为了方便读者学习，本书还附赠了视频讲解、关键词等资源，感兴趣的读者可扫描下面的二维码下载。

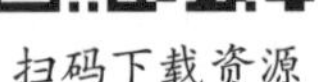

扫码下载资源　　扫码观看视频

致谢

衷心感谢王晓铃女士对本书编写工作所提供的帮助。

感谢所有阅读本书的读者，敬请你对本书提出合理化建议。

目 录 CONTENTS

基础篇

实践篇

12/25
25

后期制作篇

基础篇

第1章　AI摄影简介

随着AI技术的不断发展，越来越多的绘画软件横空出世。通过简单的指令提示，就能让AI在短时间内创作出相应的作品，除了绘画领域，摄影行业同样因AI的发展变得简单、便捷。

1.1 AI摄影的诞生和优势

摄影又称摄影术，即通过相机镜头将反射于物体上的光线在感光材料上感光并形成影像的过程。目前，摄影逐渐成为人们工作、学习和生活娱乐中必不可少的一部分。

AI技术的不断发展，让摄影行业同样变得简单、便捷。利用AI进行摄影作品的创作，既可以为摄影师提供创作灵感、改善图像质量，还可以降低被拍摄者的成本，足不出户就可以收获想要的照片。AI为摄影创作带来了全新的可能。

1.2 常用的AI摄影软件

目前市面上的AI工具有很多，下面简单介绍几种常用的AI摄影软件。

- Midjourney

Midjourney是搭载在Discord上的一款AI绘画软件，于2022年3月面向公众发布，同样也可以用于摄影作品的创作。Midjourney的使用对于新手十分友好，操作简单便捷，只需输入指令即可生成作品，易上手且出图效果较好。

- Stable Diffusion

Stable Diffusion是2022年由Stability AI公司发布的从文本生成图像的模型，是目前市场上比较热门的AI绘画工具之一，同样也可以生成摄影作品。Stable Diffusion可以免费使用，但对硬件配置要求较高。

- 文心一格

文心一格是百度集团旗下的一款AI产品，是基于文心大模型能力的AI艺术和创意辅助平台。只需在指定位置描述自己想要的照片，并选择照片对应的风格，就能根据数据模型快速生成一幅或多幅不同风格的创意作品。文心一格对中文用户的语义理解更到位，适合中文环境下的使用。

- Playground AI

Playground AI是一款简单易用、适合零基础用户的AI绘画工具。用户每天可以免费生成1000张图，通过简单的页面操作就可以轻松生成贴合提示词的高质量作品。

- Vega AI

Vega AI是国内初创公司右脑科技推出的AI绘画创作平台，是一款免费的在线AI绘画工具，支持训练AI绘画模型、文生图、图生图、姿势生图等多种创作模式。该平台界面简洁，操作简单，适合初学者迅速上手。

6pen Art

6pen Art 是国内面包多团队研发的一款 AI 绘画软件。该软件支持中英文模式，输入简单的提示词就能得到图像。用户可以自由调配模型、参考图、分辨率、艺术家等，以达到更好的出图效果。

第2章　摄影的基本知识

想要学习摄影、创造好看的照片，必须学习相关的摄影技巧，加上大量的应用与实践，这样才能在摄影的基础上利用 AI 进行创作。

2.1 焦距控制

焦距是指相机镜头的光学中心到被拍摄物体之间的距离。不同的焦距在画面中呈现的视角效果不同。数值越小，焦距越短，视角则越广；数值越大，焦距越长，视角则越窄。例如，300mm 焦距的镜头，视角只有 8°；而 8mm 焦距的超广角镜头，视角则可以达到 180°。

长焦距压缩视野，放大远处的物体；短焦距扩展视野，呈现更广阔的画面。人眼视角下的画面与 50mm 焦距的镜头拍摄的效果最接近，因此 50mm 被定义为标准焦距。广角适合拍摄风景、建筑等；标准焦距适合拍摄街景、人像等；长焦距适合拍摄远处的细节。通常情况下，焦距在 24mm 以下的镜头，适合拍摄风光类题材，如图 2.1–1 所示；而焦距为 50mm、85mm 的镜头，适合拍摄人像，如图 2.1–2 和图 2.1–3 所示。

图 2.1–1

图 2.1–2

图 2.1–3

2.2 曝光控制

曝光可以简单地理解为图像的明暗程度，曝光不足会导致照片太暗；曝光过度会导致照片太亮。而画面的曝光程度由三个参数共同决定，即感光度（ISO）、光圈、快门速度，合称为曝光三要素。通过控制这三个参数，就可以自行调节曝光量，从而实现各种复杂拍摄环境下的正确曝光。

2.2.1 感光度

感光度用于衡量底片对于光的灵敏程度。感光度的等级呈倍数关系，常见的有 ISO 100、200、400、800、1600。数值每提高一倍，感光的速率也提高一倍。

- 低感光度：ISO 800 以下

画面效果平滑、细腻。通常情况下，只要画面清楚，推荐使用低感光度进行拍摄。

- 中感光度：ISO 800~6400

中感光度设定可以降低手持相机拍摄的难度，并提高在低照明条件下拍摄的成功率。但相应地，画面中会出现一些噪点。

- 高感光度：ISO 6400 以上

推荐在暗光环境下拍摄，如星空摄影等，可以让画面效果明显变亮。但照片中将出现明显的噪点，导致画面效果粗糙、不细腻。

2.2.2 光圈

光圈是镜头内部的一个组件，用于控制透过镜头的光线量和照片的清晰或虚化效果，光圈值用 f 表示。f 值越小，光圈越大，进光量越多，虚化效果越好；f 值越大，光圈越小，进光量越少，虚化效果越差。

- f / 1.4 ~ f / 4

通常情况下，该范围内的光圈可以尽可能多地采集光线，非常适用于暗光室外（如拍摄夜空）或弱光室内拍摄（如在灯光昏暗的条件下拍摄婚宴或公司活动场景等）。

除此之外，大光圈还可以得到非常浅的景深，清晰范围很小，容易拍出主体模糊而背景清晰的效果，有效地将主体与背景分离。

- f / 5.6 ~ f / 8

通常情况下，该范围内的光圈可以提供最好的整体清晰度，是人像摄影、景观摄影、建筑摄影的理想范围。如果需要更大的景深，则推荐使用 f / 8 的光圈。

- f / 11 ~ f / 22

通常情况下，该范围内的光圈适用于需要高景深的场合，如风景摄影、建筑摄影和微距摄影等。但当光圈超过 f / 8 时，因为镜头衍射的影响，画面可能会失去锐度。

通常情况下，不推荐使用 f / 22 以上的光圈，这样画面的进光量很小，清晰度会受到很大影响，所以尽量避免使用。

2.2.3 快门速度

快门速度代表光线照射图像感应器的时间长短。分母越小，快门越慢，曝光时间越长，感光元件接触到的光线越多，画面效果越亮；分母越大，快门越快，曝光时间越短，感光元件接触到的光线越少，画面效果越暗。

除此之外，快门速度越快，呈现的画面越清晰，适用于拍摄高速运动的物体；反之，如果需要记录物体的运动轨迹，如灯轨、流水等，就需要降低快门速度，进行长时间的曝光。

- 1s ~ 1/15s

需要使用相机支架。适合在暗淡照明情况下，使用小光圈获得大景深和足够的曝光量。该范围的快门速度适用于拍摄无生命的物体或稳定不动的被摄主体。

- 1/30s ~ 1/60s

适用于手持相机，并配以闪光灯、标准镜头或广角镜头的拍摄情况。该范围的快门速度适用于照明条件不太理想的场景，使用小光圈以增大景深。适合拍摄多云天气、阴影处等。

- 1/125s ~ 1/250s

适合抓拍户外日光下以中等速度运动的物体，使用小光圈或中等大小的光圈，能产生很好的景深。该范围的快门速度适用于抓拍一些中等速度的动作，如走动着的人、运动场景、检阅活动等。

- 1/500s

该快门速度适用于抓拍运动速度较快的物体，如中等距离外的运动员、奔跑着的动物、行驶中的轿车等。

- 1/1000s

该快门速度适用于抓拍高速运动的物体，如赛车、飞翔中的鸟、野外和体育场内的比赛项目等。

2.3 景深控制

景深效果是指在对主体对焦后，在这个清晰的焦点前后范围内形成的清晰区域。简单来说即画面中景象清晰的范围。浅景深会导致画面中清晰的范围较小，深景深会导致画面中清晰的范围较大。

影响景深的因素有光圈的大小、焦距的长短、拍摄距离的远近。

- 光圈

光圈越大（光圈值 f 越小），背景虚化效果越好，景深越浅，多用于人像摄影；光圈越小（光圈值 f 越大），背景清晰程度越高，景深越深，多用于风景摄影。

- 焦距

相同光圈下，焦距越长，景深越浅；焦距越短，景深越深。

- 拍摄距离

相机与被摄主体之间的距离越近，景深越浅；距离越远，景深越深。

2.4 拍摄技巧

在了解基本的摄影参数后，还需要了解相关的拍摄技巧，才能辅助用户拍摄出更高质量的画面。

2.4.1 构图

构图技巧是摄影的根基，用以支撑画面的效果。下面介绍常见的几种构图方式，具体画面效果可参考第 12 章内容。

- 对角线构图

对角线构图是一种将画面中的主体或引导线沿对角线方向布局的构图方式，也是一种常用的构图技巧，可以吸引观众视线，让画面更具空间感和生动感。根据不同的景别和主题，可以选择直线型、曲线型或折线型，适用范围也比较广泛。

- 框架式构图

框架式构图通常会选取一些有框架感的前景罩住拍摄主体，常见的前景有门、窗、栏杆、树枝等，能创造遮挡感，并快速引导观众视线。在构图时，通常会注意边框的对齐平直和前景内容的选择。

- 中心构图

中心构图是指将主体放在画面中间位置的构图方式，可以形成强烈的视觉焦点，将画面的核心内容直观地展示给观众。在其他信息的烘托和呼应下，可以为被摄主体制造出良好的视觉效果。对于很多不会构图的新手来说，中心构图是最有保障的构图方式。

- 对称构图

对称构图又称均衡式构图、对等式构图，是指位于画面中垂线或水平线两侧所拍摄的物体基本对等的构图方式，可以形成布局平衡、结构规矩的画面效果，给观众带来一种稳定、正式、均衡的感受。

2.4.2 用光

摄影用光的主要因素包括光度、光位、光质与光色等。

- 光度

光度是指光线在物体表面的呈现度，受感光度、光圈和快门速度的影响。具体可参考 2.2 节内容。

- 光位

光位是指光线的方向与角度。同一对象在不同的光位下可以产生不同的明暗造型效果。常见的光位见表 2.4-1。

表 2.4-1 常见的光位

光 位	说 明
正面光	光线来自被摄主体正面，随角度高低分为平射光、顺光和高位逆光。通常情况下，正面光照射下的画面通透、明亮，但缺乏明暗变化，立体感较差
侧面光（前侧光）	光线来自被摄主体 45° 方位，是常用的摄影光位。通常情况下，侧面光照射下的画面具有明亮感和立体感
侧光	光线来自被摄主体 90° 方位。通常情况下，侧光照射下的画面呈现阴阳效果，是人像摄影中常见的光位，能强化明暗对比度
后侧光（侧逆光）	光线来自被摄主体的侧后方。通常情况下，后侧光照射下的画面会使被摄主体的一侧产生轮廓线条，从而分离主体与背景，是人像摄影中常见的光位

续表

光 位	说 明
逆光（背光）	光线来自被摄主体的正后方。通常情况下，逆光照射下的画面能使被摄主体产生完整的轮廓线条，从而使画面产生立体感和空间感
顶光	光线来自被摄主体的正上方。常见的包含正午的阳光、室内的灯光等。通常情况下，顶光会使人物脸部产生浓重阴影，不建议用于人像拍摄

● 光质

光质是指光线的基本性质：聚、散、软、硬等。

聚光来自同一个明显的方向，在被摄主体上产生的阴影明晰而浓重；而散光则来自若干个方向，在被摄主体上产生的阴影柔和且不明晰。

光的软硬程度取决于若干因素，狭窄的光束通常比宽广的光束要硬一些。硬光能使被摄主体产生强烈的明暗对比，有助于质感的表现；软光则利于展示物体的外形、形状和色彩等。

● 光色

光色又称色温，是光线中不同颜色含量的计量单位。

低色温（4000K 以下）呈现的颜色偏暖；高色温（6000K 以上）呈现的颜色较冷；中间色温（4000 ~ 6000K）即中性偏白光。红色是色温值最低的颜色，其次为橙、黄、白，色温最高的颜色是蓝色。通常将红、橙、黄等颜色归为暖色调，将青、蓝、白等颜色归为冷色调。

因为在自然界的大多数环境下，人们肉眼看到的物体颜色都会随着不同的光线而呈现出颜色偏差。而相机中常用的白平衡参数，即相机在不同光源下，将白色物体还原为白色的功能。如果想要拍摄的照片效果偏冷色调，则可以将白平衡数值的设置调低；如果想要拍摄的照片效果偏暖色调，则可以将白平衡数值的设置调高。

实践篇

第3章　自然风景摄影：通过AI看世界

自然风景摄影，是指以展现自然风光之美为主要创作题材的原创作品，其中包含了自然景色摄影、城市建筑摄影等，是摄影中一个常见的门类。自然风景摄影可以给人带来全面的美的享受，给予感官和心灵的愉悦。

3.1 春日风景

春日风景效果图如图 3.1–1 所示。

图 3.1–1

主题：春日风景　　软件：Midjourney

光圈值：f / 5.6　　感光度：ISO 100　　焦距：35mm

色彩：暖色为主　　光影：环境光　　构图：远景

提示词：Photograph, white bus in middle of big green field in the countryside, in the style of neo–geo minimalism, spectacular backdrops, 32K, precise ISO 100 settings, shot on 35mm, f / 5.6, ––ar 3:4 ––v 6.0

照片，白色巴士在乡村绿色的田野中间，新地理极简主义的风格，壮观的背景，32K，精确的 ISO 100 设置，焦距 35mm，光圈值 f / 5.6，出图比例 3 ：4，版本 v 6.0

其他模板：

模板①提示词：Photography, spring, Sony camera, realistic style, a train going through green leaves, traditional landscapes, green and pink, UHD image, precise ISO 150 settings, shot on 100mm, f / 11,––ar 2:3 ––v 6.0

摄影，春天，索尼相机，写实风格，一列穿过绿叶的火车，传统风景，绿色和粉色，高清图像，精确的 ISO 150 设置，焦距 100mm，光圈值 f/ 11，出图比例 2：3，版本 v 6.0

模板②提示词：Photography, spring, a picture of a street with blossoming cherry trees, ethereal clouds, sunshine, Canon camera, realistic style, UHD image, precise ISO 100 settings, shot on 35mm, f/ 2.8, --ar 3:4 --v 6.0

摄影，春天，街道上盛开的樱花树，飘逸的云，阳光，佳能相机，写实风格，高清图像，精确的 ISO 100 设置，焦距 35mm，光圈值 f/ 2.8，出图比例 3：4，版本 v 6.0

例：替换成模板①，如图 3.1–2 所示。

图 3.1–2

Tips

春日风景摄影的最大亮点便是呈现自然环境中丰富多样的颜色，如花的颜色、天空的颜色、拍摄主体的颜色。如果能够搭配好这些颜色，就能让画面呈现出更具质感和冲击力的效果。但同样地，画面中的颜色不是越多越好。建议遵循简洁的原则，一张照片中的颜色尽量不要超过 4 种，否则画面会显得很杂乱。

3.2 夏日风景

夏日风景效果图如图 3.2-1 所示。

图 3.2-1

主题：夏日风景　　软件：Midjourney

光圈值：f / 8.0　　感光度：ISO 150　　焦距：50mm

色彩：暖色为主　　光影：环境光　　构图：中景

提示词：A photo of the park in front of sea, summer, a red house in the distance, surrounded by white fences and green grass, with trees and daisies blooming on both sides, blue sky, clear details, high resolution, with natural light, bright colors, impressionism, 8K, precise ISO 150 settings, shot on 50mm, f / 8.0, --ar 3:4 --v 6.0

海景前的公园照片，夏天，远处的红房子，四周是白色的栅栏和绿色的草地，两边是树木和盛开的雏菊，天空湛蓝，细节清晰，分辨率高，带有自然光，色彩鲜艳，印象派风格，8K，精确的 ISO 150 设置，焦距 50mm，光圈值 f / 8.0，出图比例 3：4，版本 v 6.0

其他模板：

模板①提示词：Photography, summer, floating out the window of the summer, blue days and green trees, UHD image, precise ISO 150 settings, shot on 50mm, f / 11,--ar 3:2 --v 6.0

摄影，夏天，飘窗外的夏日，蔚蓝的天和绿树，高清图像，精确的 ISO 150 设置，焦距 50mm，光圈值 f / 11，出图比例 3：2，版本 v 6.0

模板②提示词：Photography, some people and shops on the road, summer, the most beautiful seaside promenade, lined with green trees and white marble walls on both sides of the road, blue sky above shines brightly, in front of it lies a vast sea, surrounded by distant mountains under the sunlight, precise ISO 150 settings, shot on 35mm, f / 11, --ar 3:2 --v 6.0

摄影，路上有一些人和商店，夏天，最美丽的海滨长廊，道路两侧是绿色的树木和白色的大理石墙，蔚蓝的天空明亮地照耀着，前面是一片广阔的海洋，周围是阳光下的远山，精确的 ISO 150 设置，焦距 35mm，光圈值 f / 11，出图比例 3：2，版本 v 6.0

例：替换成模板①，如图 3.2–2 所示。

图 3.2–2

Tips

由于夏天光线强、气温高，因此在进行自然风景摄影时，推荐采用较高的快门速度和较小的光圈，不用担心焦点不实或景深不够。但需要尽可能避开光线过强的中午，或者避免在逆光下拍摄。如果拍摄中涉及人物，可以考虑对主体人物使用反光板或闪光灯等，以降低反差，并增强阴影部分的质感。

3.3 秋日风景

秋日风景效果图如图 3.3–1 所示。

图 3.3–1

主题：秋日风景	软件：Midjourney	
光圈值：f / 2.8	感光度：ISO 200	焦距：100mm
色彩：暖色为主	光影：环境光	构图：中景

提示词：Photograph, a sparkling stream, rocks on the shore, sunsets, mists, autumn leaves, landscape photo, highly detailed, high resolution, ultra–realistic, photo realistic, 8K, precise ISO 200 settings, shot on 100mm, f / 2.8, --ar 3:2 --v 6.0

照片，波光粼粼的溪流，岸边的岩石，日落，薄雾，秋叶，风景照片，非常细节，高分辨率，超逼真，逼真的照片，8K，精确的 ISO 200 设置，焦距 100mm，光圈值 f / 2.8，出图比例 3∶2，版本 v 6.0

其他模板：

模板①提示词：Photography, maple road, Canada, the car flying on the road, autumn, realistic, UHD image, the bird's eye view, precise ISO 100 settings, f / 11, --ar 3:4 --v 6.0

摄影，枫叶大道，加拿大，有车在公路上奔驰，秋天，写实主义，超高清图像，鸟瞰视角，精确的 ISO 100 设置，光圈值 f / 11，出图比例 3∶4，版本 v 6.0

模板②提示词：Landscape photography, people boating on the lake, early morning autumn, autumn shadows by the lake, sunlight filtering through the autumn leaves, golden tones, clouds, natural light, 4K, precise ISO 100 settings, shot on 100mm, f / 8.0, --ar 16:9 --v 6.0

风景摄影，湖上有划船的人们，秋日的清晨，湖边秋影，阳光透过秋叶，金色色调，云彩，自然光，4K，精确的 ISO 100 设置，焦距 100mm，光圈值 f / 8.0，出图比例 16∶9，版本 v 6.0

例：替换成模板①，如图 3.3–2 所示。

图 3.3–2

Tips

人们通常会选择枫叶来代表秋天，而对枫叶的拍摄，重点在于体现光影，这样才能体现色彩和生命力。光线的应用是有讲究的。例如，点测光最好配合逆光拍摄，可以发现逆光的条件下更能体现树叶的通透感。除此之外，还可以利用曝光补偿来平衡画面的亮度。

3.4 冬日风景

冬日风景效果图如图 3.4–1 所示。

图 3.4–1

主题：冬日风景	软件：Midjourney	
光圈值：f / 8.0	感光度：ISO 100	焦距：70mm
色彩：暖色为主	光影：环境光	构图：中景

提示词：Photo, a farm on a snow filled path, light gold and orange, soft, romantic landscapes, tightly composed scenes, serene and tranquil scenes, light–filled scenes, pastoral charm, 8K, precise ISO 100 settings, shot on 70mm, f / 8.0, ––ar 2:3 ––v 6.0

摄影，一个在白雪覆盖的小路上的农场，淡金色和橙色，柔和，浪漫的风景，紧密组成的场景，宁静的场景，充满光线的场景，田园的魅力，8K，精确的 ISO 100 设置，焦距 70mm，光圈值 f / 8.0，出图比例 2：3，版本 v 6.0

其他模板：

模板①提示词：Photography, snowman, snow under the street lights, precise ISO 150 settings, shot on 85mm, f / 8.0, --ar 2:3 --v 6.0

摄影，雪人，街灯下的雪，精确的 ISO 150 设置，焦距 85mm，光圈值 f / 8.0，出图比例 2：3，版本 v 6.0

模板②提示词：Landscape photography, winter, a photo of a snow covered village river valley, lights, tranquil, white and brown, 4K, precise ISO 150 settings, shot on 35mm, f / 8.0, --ar 16:9 --v 6.0

风景摄影，冬天，一张白雪覆盖的村庄河谷的照片，灯光，宁静，白色和棕色，4K，精确的 ISO 150 设置，焦距 35mm，光圈值 f / 8.0，出图比例 16：9，版本 v 6.0

例：替换成模板①，如图 3.4-2 所示。

图 3.4-2

Tips

雪是冬日里一道常见的风景，但全是雪的景色看上去会给人沉闷之感，因此可以在画面中增加一些有色彩的物体作为前景或背景，以增加空间深度和画面亮点，如铺着积雪的树枝、橙色光亮的灯杆，或者花花绿绿的建筑物等。这样整个画面的内涵才更加丰富，不至于因为画面的寡淡而使观者产生厌倦的情绪。

3.5 雪山风景

雪山风景效果图如图 3.5-1 所示。

图 3.5-1

主题：雪山风景	软件：Midjourney	
光圈值：f / 11	感光度：ISO 200	焦距：50mm
色彩：冷色为主	光影：环境光	构图：全景

提示词：Photo, a mountain range surrounded by snow and an ice lake, in the style of realistic fantasy artwork, flat backgrounds, 8K, precise ISO 200 settings, shot on 50mm, f / 11, --ar 2:3 --v 6.0

照片，一座被雪和冰湖包围的山脉，现实主义的幻想艺术风格，平坦的背景，8K，精确的 ISO 200 设置，焦距 50mm，光圈值 f / 11，出图比例 2∶3，版本 v 6.0

其他模板：

模板①提示词：A picturesque autumn scene featuring the mountain in Canada, reflected on water with lush greenery and colorful leaves, natural beauty photography, the serene lake reflects the majestic mountains under clear blue skies, high-resolution photography captures every detail of nature's grandeur, in the style of impressionist, precise ISO 150 settings, shot on 35mm, f / 8.0, --ar 5:3 --v 6.0

以加拿大的山脉为特色的风景如画的秋景，水面反射着葱郁的绿色植物和五颜六色的树叶，自然美景摄影，宁静的湖水在清澈的蓝天下映衬着雄伟的山脉，高分辨率摄影捕捉到大自然宏伟的每一个细节，印象派风格，精确的 ISO 150 设置，焦距 35mm，光圈值 f / 8.0，出图比例 5∶3，版本 v 6.0

模板②提示词：Landscape photography, a picture of a beach and snow mountains, in the style of mountainous vistas, dark black and light azure, eroded surfaces, highly detailed environments, 4K, precise ISO 200 settings, shot on 50mm, f / 8.0, --ar 2:3 --v 5.2

风景摄影，一张沙滩和雪山的照片，山地景观的风格，深黑色和浅蓝色，侵蚀表面，高度详细的环境，4K，精确的 ISO 200 设置，焦距 50mm，光圈值 f / 8.0，出图比例 2：3，版本 v 5.2

例：替换成模板①，如图 3.5–2 所示。

图 3.5–2

Tips

在进行雪山风景拍摄时，因为处于天寒地冻的环境，所以要提前做好相机的电池续航工作。拍摄之前，可以先将电池放在防寒衣物的内兜里，等到拍摄时再拿出来。如果掉电过快，可以放在衣物里给电池保暖升温，然后进行拍摄。除此之外，还要做好防水工作，避免将相机长时间直接放在雪上，否则会进水或受潮，影响后续的使用。

3.6 日出风景

日出风景效果图如图 3.6–1 所示。

图 3.6–1

主题：日出风景　　软件：Midjourney
光圈值：f / 5.6　　感光度：ISO 200　　焦距：70mm
色彩：暖色为主　　光影：环境光　　构图：中景

提示词：Photo, sun, sunrise, sea sunrise, waves, sea, coastline, beach, 16K, precise ISO 200 settings, shot on 70mm, f / 5.6, --ar 2:3 --v 6.0

照片，太阳，日出，海上日出，海浪，大海，海岸线，海滩，16K，精确的 ISO 200 设置，焦距 70mm，光圈值 f / 5.6，出图比例 2∶3，版本 v 6.0

其他模板：

模板①提示词：Photography, sun, sunrise, a cabin in the snow covered with snow on top, richly colored sky, orange and azure, the snapshot aesthetic, 32K, precise ISO 150 settings, shot on 70mm, f / 5.6, --ar 4:3 --v 6.0

摄影，太阳，日出，雪中的小屋、屋顶堆满白雪，色彩丰富的天空，橙色和蔚蓝色，快照美学，32K，精确的 ISO 150 设置，焦距 70mm，光圈值 f / 5.6，出图比例 4∶3，版本 v 6.0

模板②提示词：Landscape photography, reflections of Chinese buildings in the river at sunrise, in the style of photo-realistic techniques, mesmerizing colors, 4K, precise ISO 150 settings, shot on 50mm, f / 8.0, --ar 16:9 --v 6.0

风景摄影，日出时中式建筑在河中的倒影，逼真的技术风格，迷人的色彩，4K，精确的 ISO 150 设置，焦距 50mm，光圈值 f / 8.0，出图比例 16∶9，版本 v 6.0

例：替换成模板①，如图 3.6-2 所示。

图 3.6-2

Tips

不同季节、不同时间段的日出的表现都不一样，所以在拍摄时，必须重视时间的选择。拍摄日出的最佳季节是春、秋两季。这两季的日出比夏季的日出晚，并且日落早，对拍摄有利，而且云层较多，比较容易遇到“彩霞满天”的情景。

3.7 峡谷风景

峡谷风景效果图如图 3.7–1 所示。

图 3.7–1

主题：峡谷风景　软件：Midjourney

光圈值：f / 5.6　感光度：ISO 200　焦距：70mm

色彩：暖色为主　光影：环境光　构图：全景

提示词：Real photography, canyon, someone is boating on the stream, river water, cliff nature, cave, mountain photography, realistic depiction of light, light navy and light amber, 32K, precise ISO 200 settings, shot on 70mm, f / 5.6, --ar 2:3 --v 6.0

真实的摄影，峡谷，一个人在小溪上划船，河水，自然的崖壁，洞穴，山景摄影，写实描绘光，浅海军蓝和浅琥珀色，32K，精确的 ISO 200 设置，焦距 70mm，光圈值 f / 5.6，出图比例 2∶3，版本 v 6.0

其他模板：

模板①提示词：Photography, canyon, a beautiful waterfall, epic fantasy scenes, backlight, 8K, precise ISO 400 settings, shot on 70mm, f / 8.0, --ar 2:3 --v 6.0

摄影，峡谷，一条美丽的瀑布，史诗般的幻想场景，背光，8K，精确的 ISO 400 设置，焦距 70mm，光圈值 f / 8.0，出图比例 2∶3，版本 v 6.0

模板②提示词：Real photography, Yarlung Zangbo Grand Canyon, blue sky, eye-catching composition, 32K, precise ISO 100 settings, f / 8.0, --ar 3:4 --v 6.0

真实的摄影，雅鲁藏布大峡谷，蓝天，引人注目的构图，32K，精确的 ISO 100 设置，光圈值 f / 8.0，出图比例 3∶4，版本 v 6.0

例：替换成模板①，如图 3.7-2 所示。

图 3.7-2

Tips

一般情况下，每天中午 11 点至下午 1 点是峡谷的最佳拍摄时间，此时光线会从峡谷上方照射进来，给峡谷以充分照明。而在捕捉光线的同时，要记住高对比感对画面的影响。可以在最暗的地区和最亮的地区间找到许多具有高反差的场景进行拍摄。

3.8 瀑布风景

瀑布风景效果图如图 3.8-1 所示。

图 3.8-1

主题：瀑布风景　　软件：Midjourney
光圈值：f / 11　　感光度：ISO 200　　焦距：35mm
色彩：暖色为主　　光影：环境光　　构图：远景

提示词：Photo, the Iguazu Falls in Brazil, seen from above, the water falls down, sunshine, the yellow flow is surrounded by green vegetation on top of rocks, in front there's blue sky with white clouds, aerial view, high resolution, 8K, precise ISO 200 settings, shot on 35mm, f / 11, --ar 3:2 --v 6.0

照片，巴西的伊瓜苏瀑布，从上俯视，落下来的水，阳光，黄色的水流被岩石顶部的绿色植被所包围，前面是蓝天白云，鸟瞰图，高分辨率，8K，精确的 ISO 200 设置，焦距 35mm，光圈值 f / 11，出图比例 3：2，版本 v 6.0

其他模板：

模板①提示词：Photography, aerial view of the Iguazu Falls in Brazil, the water falls down to an abyss below, aerial photography, drone footage, high resolution, water cascades over mosscovered rocks and into chasms, water mist rises from flowing white waters, 8K, precise ISO 150 settings, shot on 35mm, f / 11, --ar 3:2 --v 6.0

摄影，巴西伊瓜苏瀑布鸟瞰图，水流至下方的深渊，摄像机位于上方，航拍，无人机镜头，高分辨率，水从长满苔藓的岩石上倾泻而下，水雾从流动的白色水域升起，8K，精确的 ISO 150 设置，焦距 35mm，光圈值 f / 11，出图比例 3：2，版本 v 6.0

模板②提示词：Photo of falls, golden yellow water falls from the top to the bottom, the waterfall is surrounded by lush green vegetation and a blue sky with white clouds, sunlight reflects on the surface creating a misty effect around the fall, high angle view, 4K, precise ISO 150 settings, shot on 50mm, f / 8.0, --ar 16:9 --v 6.0

瀑布的照片，金黄色的水从顶部落到底部，瀑布周围是郁郁葱葱的绿色植被和蓝天白云，反射在地面上的阳光在瀑布周围形成薄雾效果，高视角，4K，精确的 ISO 150 设置，焦距 50mm，光圈值 f / 8.0，出图比例 16：9，版本 v 6.0

例：替换成模板①，如图 3.8-2 所示。

图 3.8-2

Tips

瀑布本身有自己独特的轮廓和形状，可以为画面打造出线条构图的效果。但线条也有横、竖、曲、直之分，不同的线条可以表现出不同的画面效果。例如，横向线条（水平线）能展现出大场景；竖向线条则以向上的形式引导视线；斜线，尤其是对角线，则具有流动感；弯曲的线条（C弧线和S形线）则带着流动感、流畅感。用户可以根据瀑布和拍摄位置的具体情况来选择构图方式。

3.9 烟花风景

烟花风景效果图如图 3.9–1 所示。

图 3.9–1

主题：烟花风景	软件：Midjourney	
光圈值：f / 8.0	感光度：ISO 200	焦距：70mm
色彩：暖色为主	光影：环境光	构图：中景

提示词：Photo, firework, ancient Chinese architecture, 16K, precise ISO 200 settings, shot on 70mm, f / 8.0, --ar 3:2 --v 6.0

照片，烟花，中式建筑，16K，精确的 ISO 200 设置，焦距 70mm，光圈值 f / 8.0，出图比例 3：2，版本 v 6.0

其他模板：

模板①提示词：Real photo shoot, lots of colorful fireworks,fireworks on the river, lamplight, bright, Sony camera, 16K, precise ISO 200 settings, shot on 85mm, f / 8.0, --ar 3:2 --v 6.0

真实的照片拍摄，五颜六色的烟花，河上的烟花，灯光，明亮，索尼相机拍摄，16K，精确的 ISO 200 设置，焦距 85mm，光圈值 f / 8.0，出图比例 3：2，版本 v 6.0

模板②提示词：Landscape photography, fireworks on a yellow background with sparklers, dark cyan and orange, sparkle, metalworking mastery, 4K, precise ISO 600 settings, shot on 70mm, f / 5.6, --ar 2:3 --v 6.0

风景摄影，黄色背景中的烟花与火花，深青色和橙色，闪光，金属质地，4K，精确的 ISO 600 设置，焦距 70mm，光圈值 f / 5.6，出图比例 2：3，版本 v 6.0

例：替换成模板①，如图 3.9-2 所示。

图 3.9-2

Tips

在拍摄烟花的过程中，由于需要长时间曝光，推荐使用较低的 ISO 数值，通常情况下设置为100~300即可。长时间曝光会让画面呈现更多的噪点，所以使用高的ISO数值会严重影响画质。但如果是在距离烟花位置较远、周围环境光线较弱的情况下，使用太低的 ISO 数值进行拍摄，会导致画面曝光严重不足，这时可以适当提高 ISO 数值。

3.10 星空风景

星空风景效果图如图 3.10-1 所示。

图 3.10-1

主题：星空风景　**软件：**Midjourney
光圈值：f / 2.8　**感光度：**ISO 1600　**焦距：**24mm
色彩：暖色为主　**光影：**环境光　**构图：**远景

提示词：Photograph, starry skies on the mountains, a variety of bright colors, 8K, precise ISO 1600 settings, shot on 24mm, f / 2.8, --ar 3:2 --v 6.0

照片，山上的星空，各种鲜艳的色彩，8K，精确的 ISO 1600 设置，焦距 24mm，光圈值 f / 2.8，出图比例 3：2，版本 v 6.0

其他模板：

模板①提示词：Photography, starry sky, footpath, someone walking on the road, bright multiple color, 16K, precise ISO 1600 settings, shot on 35mm, f / 2.8, --ar 2:3 --v 6.0

摄影，星空，步道，一个人走在路上，鲜艳的多重色彩，16K，精确的 ISO 1600 设置，焦距 35mm，光圈值 f / 2.8，出图比例 2：3，版本 v 6.0

模板②提示词：Photo, starry sky, people boating on the sea, night, bright color, 8K, precise ISO 1600 settings, shot on 35mm, f / 2.8, --ar 3:2 --v 6.0

照片，星空，人们在海上泛舟，夜晚，鲜艳的色彩，8K，精确的 ISO 1600 设置，焦距 35mm，光圈值 f / 2.8，出图比例 3：2，版本 v 6.0

例：替换成模板①，如图 3.10-2 所示。

图 3.10-2

Tips

在星空摄影中，可能会遇到对焦的问题。在星星不够亮的情况下，相机无法自动对焦，这时就需要手动对焦。通常先将焦点设置在无穷远处，再根据镜头的具体情况回退一点位置。还可以采用实时取景的方式，将镜头对准夜空中最明亮的地方，然后微调对焦环，以星点细小为佳。

3.11 大海风景

大海风景效果图如图 3.11-1 所示。

图 3.11-1

主题：大海风景	软件：Midjourney	
光圈值：f / 5.6	感光度：ISO 200	焦距：24mm
色彩：暖色和冷色相结合	光影：环境光	构图：远景

提示词：A man was walking along the shore, beach scenes, blue sky, wide-angle lens, minimalism, landscape, naturalism, waves, in the style of coastal scenery, aerial photography, soft, romantic scenes, high quality, precise ISO 200 settings, shot on 24mm, f / 5.6, --ar 2:3 --v 6.0

一个沿着海岸散步的人，海滩的场景，蓝天，广角镜头，极简主义，风景，自然主义，海浪，海岸风景的风格，航拍，柔和，浪漫的场景，高品质，精确的 ISO 200 设置，焦距 24mm，光圈值 f / 5.6，出图比例 2：3，版本 v 6.0

其他模板：

模板①提示词：Photography, lighthouse in sunrise over ocean, 32K, precise ISO 600 settings, shot on 70mm, f / 11, --ar 3:2 --v 6.0

摄影，海上日出下的灯塔，32K，精确的 ISO 600 设置，焦距 70mm，光圈值 f / 11，出图比例 3：2，版本 v 6.0

模板②提示词：Landscape photography, blue sea and sky, reflection of the sun on the sea, sparkling waves, 4K, precise ISO 150 settings, shot on 50mm, f / 8.0, --ar 3:4 --v 6.0

风景摄影，蓝色的大海和天空，太阳在海面上反射，波光粼粼的波浪，4K，精确的 ISO 150 设置，焦距 50mm，光圈值 f / 8.0，出图比例 3：4，版本 v 6.0

例：替换成模板①，如图 3.11-2 所示。

图 3.11-2

Tips

在拍摄海岸四周时，可以利用小路或栏杆等物体作为“引导线”，把目光带到画面的主体上。除此之外，还可以通过对比的方式凸显大海的宏伟与广阔。例如，在画面中加入人物背影或一个常见的小屋、灯塔等建筑。

3.12 极光风景

极光风景效果图如图 3.12-1 所示。

图 3.12-1

主题：极光风景　　软件：Midjourney
光圈值：f / 2.8　　感光度：ISO 1600　　焦距：70mm
色彩：暖色为主　　光影：环境光、灯光　　构图：中景

提示词：Real photography, auroras, snow, arctic glaciers, red houses in the distance, lights, 16K, precise ISO 1600 settings, shot on 70mm, f / 2.8, --ar 2:3 --v 6.0

真实的摄影，极光，雪，北极冰川，远处的红房子，灯光，16K，精确的 ISO 1600 设置，焦距 70mm，光圈值 f / 2.8，出图比例 2：3，版本 v 6.0

其他模板：

模板①提示词：Image of a van parked in the snow, under the aurora sky in Iceland, in the style of realistic, hyper-detailed renderings, futuristic cyberpunk, sparkling water reflections, outdoor scenes, 8K, precise ISO 1600 settings, shot on 50mm, f / 2.8, --ar 3:2 --v 6.0

一辆面包车停在雪地里的图像，在冰岛的极光天空下，现实主义，超细节渲染，未来主义的赛博朋克风格，波光粼粼的水面反射，户外场景，8K，精确的 ISO 1600 设置，焦距 50mm，光圈值 f / 2.8，出图比例 3：2，版本 v 6.0

模板②提示词：Photography, aurora, polar lights in the sky, snow, dark teal and light red, forest, shadows of the forest, UHD image, precise ISO 1600 settings, shot on 50mm, f / 2.8, --ar 3:4 --v 6.0

摄影，极光，天空中的极光，雪，深青色和浅红色，森林，森林的阴影，超高清图像，精确的 ISO 1600 设置，焦距 50mm，光圈值 f / 2.8，出图比例 3：4，版本 v 6.0

例：替换成模板①，如图 3.12-2 所示。

图 3.12-2

Tips

拍摄极光需要良好的构图，如果只有极光，没有周遭的环境作为前景，整个画面就显得很单调。因此，在构图时可以找一些独特的风景作为前景，如房屋、汽车、森林等，会让画面的层次显得更丰富。除此之外，在黑暗的地方进行拍摄，光圈和 ISO 的设置同样很重要。尽量使用镜头的最大光圈，ISO 建议设置在 800~1600，因为如果设置得太低，会导致画面的亮度不够；如果设置得太高，噪点又会很多。具体参数建议结合当时的环境进行调整。

3.13 街头风景

街头风景效果图如图 3.13–1 所示。

图 3.13–1

主题：街头风景 **软件：Midjourney**

光圈值：f / 11 **感光度：ISO 200** **焦距：70mm**

色彩：暖色为主 **光影：环境光** **构图：中景**

提示词：Photo, a flower shop on the old city street, street scenes, soft and dreamy atmosphere, light–filled outdoor scenes, side view, bright, 16K, precise ISO 200 settings, shot on 70mm, f / 11, ––ar 2:3 ––v 6.0

照片，旧城街道上的一家花店，街景，柔和梦幻的氛围，光线充足的户外场景，侧视图，明亮，16K，精确的 ISO 200 设置，焦距 70mm，光圈值 f / 11，出图比例 2∶3，版本 v 6.0

其他模板：

模板①提示词：A photo shows the sun setting over an urban street scene that appears to be in France, focus on a man riding his bike through traffic, cars and people walking alongside him, buildings and trees with yellow light coming from behind, golden hour time, the sky being orange and pink, more cityscape, 8K, precise ISO 200 settings, shot on 50mm, f / 11, ––ar 2:3 ––v 6.0

一张夕阳下法国城市街道的照片，骑着自行车穿过车流的男子是画面的焦点，旁边有汽车和行人，建筑物和树木后面有黄色的光，黄金时间，天空是橙色和粉红色的，更多的城

市景观，8K，精确的 ISO 200 设置，焦距 50mm，光圈值 f / 11，出图比例 2∶3，版本 v 6.0

模板②提示词：A photo of people walking on the streets in Chicago, with buildings reflections and city lights in puddles, the reflection is clear and realistic, capturing an urban atmosphere at sunset, behind is tall buildings, a busy street scene with cars passing through, in the style of a realistic street scene at sunset, 4K, precise ISO 200 settings, shot on 50mm, f / 8.0, --ar 2:3 --v 6.0

人们走在芝加哥街道上的照片，水坑里有建筑倒影和城市灯光，倒影清晰逼真，捕捉到了日落时分的城市氛围，后面是高楼大厦，车辆穿梭的繁忙街景，日落时街景的写实风格，4K，精确的 ISO 200 设置，焦距 50mm，光圈值 f / 8.0，出图比例 2∶3，版本 v 6.0

例：替换成模板①，如图 3.13-2 所示。

图 3.13-2

Tips

在拍摄街头风景时，可以尝试转换不同的角度，如常用的低角度，可以通过超低的角度营造视觉冲击力。除此之外，街上的景色也值得留意和挑选，如街头华丽的橱窗、小的摊贩、过路的行人、五颜六色的招牌等，都可以表现出街头的人文气息。画面中的“乱”与“齐”也能够营造一种视觉上的反差。

3.14 草原风景

草原风景效果图如图 3.14-1 所示。

图 3.14-1

主题：草原风景	软件：Midjourney	
光圈值：f / 5.6	感光度：ISO 150	焦距：24mm
色彩：暖色和冷色相结合	光影：环境光	构图：全景

提示词：Real photography, image of landscape, steppe and mountain, cattle graze in the pasture, in the style of village, yellow and aquamarine, 8K, precise ISO 150 settings, shot on 24mm, f / 5.6, --ar 2:3 --v 6.0

真实的摄影，风景照，草原与高山，牛群在牧场上吃草，乡村的风格，黄色和宝石蓝，8K，精确的 ISO 150 设置，焦距 24mm，光圈值 f / 5.6，出图比例 2∶3，版本 v 6.0

其他模板：

模板①提示词：Photography, image of landscape, steppe and mountain, yellow and green, orientalist landscapes, cottage on the prairie, rustic style, forest, blue sky, 8K, precise ISO 200 settings, shot on 35mm, f / 5.6, --ar 2:3 --v 6.0

摄影，风景照，草原与高山，黄色和绿色，东方主义景观，草原上的小屋，乡村风格，森林，蓝天，8K，精确的 ISO 200 设置，焦距 35mm，光圈值 f / 5.6，出图比例 2∶3，版本 v 6.0

模板②提示词：Real photography, image of sunset over a vast unpopulated prairie, green and emerald, eye-catching composition, 32K, precise ISO 600 settings, f / 5.6, --ar 3:4 --v 6.0

真实的摄影，在一个广阔无人的草原上的日落图像，绿色和翡翠色，引人注目的构图，32K，精确的 ISO 600 设置，光圈值 f / 5.6，出图比例 3∶4，版本 v 6.0

例：替换成模板①，如图 3.14-2 所示。

图 3.14-2

Tips

在草原风景的画面元素中，最常见的就是蓝天、白云、山峦和草坪等。一般情况下，三分法构图适合大部分的草原风景摄影：1/3 天空，1/3 树木、房屋，1/3 地面。但这套方法并不适用于所有场景，不能刻板地进行套用，而是要根据具体的场景调整构图位置，将画面的重点放在风景中最出彩的地方。

3.15 沙漠风景

沙漠风景效果图如图 3.15-1 所示。

图 3.15-1

主题：沙漠风景　　软件：Midjourney

光圈值：f / 11　　感光度：ISO 100　　焦距：50mm

色彩：暖色　　光影：环境光　　构图：远景

提示词：Photo, desert, sand dunes, a man walking through the desert, 16K, precise ISO 100 settings, shot on 50mm, f / 11, --ar 3:2 --v 6.0

照片，沙漠，沙丘，一个人穿过沙漠，16K，精确的 ISO 100 设置，焦距 50mm，光圈值 f / 11，出图比例 3：2，版本 v 6.0

其他模板：

模板①提示词：Photography, sunrise in the desert, blue sky, clouds, camels in the desert, 8K, precise ISO 150 settings, shot on 70mm, f / 8.0, --ar 2:3 --v 6.0

摄影，沙漠里的日出，蓝天，云彩，沙漠里的骆驼，8K，精确的 ISO 150 设置，焦距 70mm，光圈值 f / 8.0，出图比例 2：3，版本 v 6.0

模板②提示词：Landscape photography, desert, wind, blowing sand, camel, sunset, 4K, precise ISO 100 settings, shot on 50mm, f / 11, --ar 2:3 --v 6.0

风景摄影，沙漠，风，飞扬的风沙，骆驼，落日，4K，精确的 ISO 100 设置，焦距 50mm，光圈值 f / 11，出图比例 2：3，版本 v 6.0

例：替换成模板①，如图 3.15-2 所示。

图 3.15-2

Tips

在拍摄沙漠时，如果身处风沙和灰尘较大的环境下，尽量不要更换镜头，以免沙尘进入反光镜，这有可能导致磨损加剧甚至相机故障。如果一定要在恶劣环境中更换镜头，可以先把机身翻过来，把卡口对着下方装卸，这样能最大限度地避免沙尘的进入。换下来的镜头应当尽快盖好前后盖，放入包内或镜头筒中。

第4章　场景摄影：还原眼中的光景

并不是必须要到名山大川才能拍出好照片，我们的日常生活中就有很多值得拍摄的场景：一座公园、一片草地、一片沙滩，这些都是我们摄影的创作灵感来源。独特的画面构图、特别的拍摄视角，可以呈现出全然不同的场景。

4.1 公园风景

公园风景效果图如图 4.1–1 所示。

图 4.1–1

主题：公园风景　　软件：Midjourney
光圈值：f / 11　　感光度：ISO 100　　焦距：50mm
色彩：暖色为主　　光影：环境光　　构图：全景

提示词：The park in autumn, a Chinese–style pavilion stands on the lake, surrounded by trees and white stone bridges, the background is a blue sky with some yellow autumn leaves, bright colors, natural light, an elegant atmosphere, in the style of Chinese landscape painting, 8K, precise ISO 100 settings, shot on 50mm, f / 11, ––ar 3:2 ––v 6.0

秋天的公园，湖上矗立着一座中式的亭子，周围是树木和白色的石桥，背景是蓝色的天空和一些黄色的秋叶，色彩鲜艳，自然光，优雅的气氛，具有中国山水画的风格，8K，精确的 ISO 100 设置，焦距 50mm，光圈值 f / 11，出图比例 3：2，版本 v 6.0

其他模板：

模板①提示词：A beautiful park, on an autumn morning, with red maple trees and milky white cypress trees, featuring vibrant colors, high definition images, bright sunshine, warm tones, UHD image, precise ISO 150 settings, shot on 50mm, f / 11,--ar 2:3 --v 6.0

一座美丽的公园，在一个秋天的早晨，有红色的枫树和乳白色的柏树，具有鲜艳的色彩，高清图像，明亮的阳光，暖色调，精确的 ISO 150 设置，焦距 50mm，光圈值 f / 11，出图比例 2：3，版本 v 6.0

模板②提示词：The park is decorated with flowers, grass and trees on both sides of the road, there is a bench in front, with colorful flower beds in the distance, a green lawn on which people can sit or lie down to rest, next to it stand stone sculptures and small fountains, with high definition and ultra details, precise ISO 200 settings, shot on 35mm, f / 11, --ar 3:4 --v 6.0

公园的道路两旁装饰着鲜花、草地和树木，前面有一条长凳，远处有五颜六色的花坛，人们可以在绿色的草坪上面坐下或躺下休息，旁边矗立着石雕和小喷泉，具有高清晰度和超细节，精确的 ISO 200 设置，焦距 35mm，光圈值 f / 11，出图比例 3：4，版本 v 6.0

例：替换成模板①，如图 4.1-2 所示。

图 4.1-2

Tips

在拍摄公园风景前，需要先了解公园的环境和特点，如公园的风格、植物、地形等，以便选择合适的拍摄主题和角度。除此之外，还要观察天气状况，选择合适的拍摄时间。例如春秋时节的公园，可以拍全景和远景；而冬天的公园大景不如小景，可以多找找角度。

4.2 城市夜景

城市风景效果图如图 4.2-1 所示。

图 4.2-1

主题：城市夜景	软件：Midjourney	
光圈值：f / 8.0	感光度：ISO 200	焦距：35mm
色彩：暖色和冷色相结合	光影：灯光	构图：远景

提示词：Photograph, night view of the Bund in Shanghai, the neon lights, the river, the reflection of the water, 8K, precise ISO 200 settings, shot on 35mm, f / 8.0, --ar 3:2 --v 6.0

照片，上海外滩的夜景，霓虹灯光，江水，水的反射，8K，精确的 ISO 200 设置，焦距 35mm，光圈值 f / 8.0，出图比例 3∶2，版本 v 6.0

其他模板：

模板①提示词：Photography, a city street at night, bus, trees, orange street lights, yellow and brown, 32K, precise ISO 150 settings, shot on 50mm, f / 8.0, --ar 2:3 --v 6.0

摄影，夜晚的城市街道，公共汽车，树木，橙色的路灯，黄色和棕色，32K，精确的 ISO 150 设置，焦距 50mm，光圈值 f / 8.0，出图比例 2∶3，版本 v 6.0

模板②提示词：Photography, city buildings at night, cityscape, night, neon lights, a stream of vehicles, precise ISO 200 settings, shot on 80mm, f / 11, --ar 3:4 --v 6.0

摄影，夜晚的城市建筑，城市景观，夜晚，霓虹灯，车流，精确的 ISO 200 设置，焦距 80mm，光圈值 f / 11，出图比例 3∶4，版本 v 6.0

例：替换成模板①，如图 4.2-2 所示。

图 4.2-2

Tips

通常情况下，夜景拍摄需要使用小光圈，有利于发挥镜头的最佳分辨率，还能增加景深效果，把前景和背景都拍得很清晰。一般来说，可以收几档光圈，但不宜太小，否则会降低成像质量。一般情况下，光圈值控制在 f / 8.0~f / 14 最佳。

4.3 江南建筑

江南建筑效果图如图 4.3-1 所示。

图 4.3-1

主题：江南建筑　　软件：Midjourney
光圈值：f / 11　　感光度：ISO 150　　焦距：35mm
色彩：冷色为主　　光影：环境光　　构图：全景

提示词：Photograph, a small village in southern China, with mountains and water reflections, the houses have white walls and red tiles, reflecting the scenery of Chinese architecture, in front is an ancient river, surrounded by green trecs and distant mountain peaks, panoramic lens, showing the landscape of a picturesque rural life in the style of southern culture, highlights traditional cultural elements such as window decorations and wooden bridges on riverside areas, 4K, precise ISO 150 settings, shot on 35mm, f / 11, --ar 3:2 --v 6.0

照片，中国南方的一个小村庄，有山和水的倒影，这些房子有白墙红瓦，反映了中国建筑的特征，前面是一条古老的河流，被绿树和远山环绕，全景镜头，展现了南方文化风格的田园生活风景，突出了如沿江地区的窗户装饰和木桥等传统文化元素，4K，精确的ISO 150 设置，焦距 35mm，光圈值 f / 11，出图比例 3：2，版本 v 6.0

其他模板：

模板①提示词：Photography, a small village in southern China, a canal that has boats on it, the wind blows through the canal, making ripples, surrounded by Chinese buildings, in the style of traditional Chinese landscape, reflection, light and shadow, natural light, symmetry, 8K, precise ISO 200 settings, shot on 50mm, f / 11, --ar 3:2 --v 6.0

摄影，中国南方的一个小村庄，一条运河上有船，风吹过运河，泛起涟漪，被中式建筑环绕，中国传统景观的风格，反射，光影效果，自然光，对称，8K，精确的 ISO 200 设置，焦距 50mm，光圈值 f / 11，出图比例 3：2，版本 v 6.0

模板②提示词：An ancient town with stone arch bridges in the style of Chinese architecture, reflections on the water surface of a small river in golden hour light, lanterns hanging from buildings on both sides, warm colors, captured with a wide angle lens, tranquil atmosphere with historical charm, 4K, precise ISO 100 settings, shot on 35mm, f / 11, --ar 3:2 --v 6.0

中国建筑风格的石拱桥古镇，黄金时间下小河水面的倒影，两岸建筑上挂着灯笼，暖色调，广角镜头捕捉，宁静的氛围和历史的魅力，4K，精确的 ISO 100 设置，焦距 35mm，光圈值 f / 11，出图比例 3：2，版本 v 6.0

例：替换成模板①，如图 4.3-2 所示。

图 4.3-2

Tips

在拍摄建筑时，想凸显建筑物的立体感，光线的把握非常重要。通常情况下，每天早上和晚上的光线较好，侧逆光或逆光可以强调建筑的质感和轮廓感；而中午的光线相对较平，无法完整地体现建筑的雕塑感，也无法体现光影效果。

4.4 欧式园林

欧式园林效果图如图 4.4–1 所示。

图 4.4–1

主题：欧式园林　　**软件：**Midjourney

光圈值：f / 8.0　　**感光度：**ISO 150　　**焦距：**35mm

色彩：暖色为主　　**光影：**环境光　　**构图：**远景

提示词：Photograph, European style courtyard, French courtyard, sunlight, angel statue, open –air, fountain, in the style of light orange and light beige, multilayered realism, high–contrast shading, detailed architecture paintings, 4K, precise ISO 150 settings, shot on 35mm, f / 8.0, ––ar 4:3 ––v 6.0

照片，欧式庭院，法式庭院，阳光，天使雕像，露天，喷泉，浅橙色和浅米色风格，多层写实，高对比阴影，细致的建筑绘画，4K，精确的 ISO 150 设置，焦距 35mm，光圈值 f / 8.0，出图比例 4∶3，版本 v 6.0

其他模板：

模板①提示词：Photography, European style garden, garden art, rose, European architecture, sunlight, lawn, precise ISO 150 settings, shot on 35mm, f / 8.0, ––ar 4:3 ––v 6.0

照片，欧式园林，园林艺术，玫瑰，欧式建筑，阳光，草坪，精确的 ISO 150 设置，焦距 35mm，光圈值 f / 8.0，出图比例 4∶3，版本 v 6.0

模板②提示词：A photo of an European style garden with ivy–covered walls, surrounded by lush green gardens and blooming rose beds, classical style, surrounded by a circular garden bed, pink roses and other flowers, blue sky, precise ISO 150 settings, shot on 35mm, f / 8.0, ––ar 3:4 ––v 6.0

一张墙上爬满了常春藤的欧式园林的照片，周围是郁郁葱葱的绿色花园和盛开的玫瑰花坛，古典风格，四周环绕着圆形的花坛，粉红色的玫瑰和其他花朵，蓝天，精确的 ISO 150 设置，焦距 35mm，光圈值 f/8.0，出图比例 3：4，版本 v 6.0

例：替换成模板①，如图 4.4-2 所示。

图 4.4-2

Tips

园林中的线条错落有致，在拍摄过程中，可以运用对角线、对称线等线条，使画面更加丰富和美观。

4.5 中式园林

中式园林效果图如图 4.5-1 所示。

图 4.5-1

主题：中式园林　　软件：Midjourney

光圈值：f / 11　　感光度：ISO 100　　焦距：50mm

色彩：暖色和冷色相结合　　光影：环境光　　构图：中景

提示词：Photograph, garden design, landscape, Chinese style, natural light, blue sky, featuring an ancient building style, stone path surface, green lawns, willow trees, bright sunshine, water reflection, tranquility, natural scenery, relaxing atmosphere, 8K, precise ISO 100 settings, shot on 50mm, f / 11, --ar 3:4 --v 6.0

照片，园林设计，景观，中式，自然光，蓝天，古建筑风格，石板小路，绿草坪，柳树，明亮的阳光，水的反射，宁静，自然风光，轻松的气氛，8K，精确的 ISO 100 设置，焦距 50mm，光圈值 f / 11，出图比例 3：4，版本 v 6.0

其他模板：

模板①提示词：Photography, a Chinese style garden with pavilions, natural light, blue sky, rockeries, the scene includes stone carvings, willow trees, grass lawns, precise ISO 150 settings, shot on 35mm, f / 11, --ar 3:2 --v 6.0

摄影，有亭台楼阁的中式园林，自然光，蓝天，假山，这个场景包括石雕、柳树和草坪，精确的 ISO 150 设置，焦距 35mm，光圈值 f / 11，出图比例 3：2，版本 v 6.0

模板②提示词：A Chinese style garden with a rockery, pond and bridge surrounded by green trees and shrubs under the blue sky, an ancient building with black tiles on its roof, in the style of traditional Chinese style, with classical elements such as eaves, columns and arched doors, precise ISO 100 settings, shot on 35mm, f / 8.0, --ar 3:4 --v 6.0

蓝天下被假山、池塘、桥、绿树、灌木环绕的中式园林，有一座古老的建筑，屋顶上有黑色的瓦，中国传统风格，采用了屋檐、柱子和拱形门等古典元素，精确的 ISO 100 设置，焦距 35mm，光圈值 f / 8.0，出图比例 3：4，版本 v 6.0

例：替换成模板①，如图 4.5-2 所示。

图 4.5-2

Tips

因为园林中的景色很多，涉及的颜色也非常多样，在摄影时可以运用色彩搭配来创造不同的氛围和效果。除此之外，还要结合园林本身的特点和意境，搭配合适的视角、光线、前景与背景等，拍出优美而生动的园林照片。

4.6 洛可可风格建筑

洛可可风格建筑效果图如图 4.6-1 所示。

图 4.6-1

主题：洛可可风格建筑	软件：Midjourney	
光圈值：f / 8.0	感光度：ISO 150	焦距：50mm
色彩：暖色为主	光影：环境光	构图：中景

提示词：Photograph, an ancient and mysterious palace hall, in the style of rococo-inspired art, realistic and hyper-detailed renderings, golden light, gigantic scale, front view, ultra wide angle, ray tracing, 16K, precise ISO 150 settings, shot on 50mm, f / 8.0, --ar 3:4 --v 6.0

照片，一个古老而神秘的宫殿大厅，洛可可风格的艺术，逼真和超细节的效果图，金色的光，巨大的规模，正面视图，超广角，光线追踪，16K，精确的 ISO 150 设置，焦距 50mm，光圈值 f / 8.0，出图比例 3：4，版本 v 6.0

其他模板：

模板①提示词：Photography, outside the castle, the castle, ancient and mysterious, in the style of rococo-inspired art, realistic and hyper-detailed renderings, ultra wide angle, precise ISO

150 settings, shot on 35mm, f / 8.0, --ar 3:4 --v 6.0

摄影，城堡外，城堡，古老而神秘，洛可可风格的艺术，逼真和超细节的效果图，超广角，精确的 ISO 150 设置，焦距 35mm，光圈值 f / 8.0，出图比例 3：4，版本 v 6.0

模板②提示词：A photo of interior design of an elegant living room, with a pastel pink and white color scheme, decorated with blooming flowers, with rococo-inspired architecture and intricate details, windows overlooking the garden with blooming cherry blossom trees, precise ISO 150 settings, shot on 50mm, f / 11, --ar 3:2 --v 6.0

一张优雅客厅的室内设计照片，柔和的粉色和白色配色，装饰着盛开的花朵，洛可可风格的建筑和复杂的细节，窗户俯瞰着盛开的樱花树花园，精确的 ISO 150 设置，焦距 50mm，光圈值 f / 11，出图比例 3：2，版本 v 6.0

例：替换成模板①，如图 4.6-2 所示。

图 4.6-2

Tips

洛可可风格建筑于 18 世纪 20 年代产生于法国并流行于欧洲，是在巴洛克式建筑的基础上发展起来的风格流派。洛可可风格的基本特点是纤弱娇媚、华丽精巧、甜腻温柔、纷繁琐细。在进行拍摄时，可以利用自然光，着重于表现其烦琐的装饰。在光线不够通透的情况下，还可以利用室内的灯光。

4.7 城市剪影

城市剪影效果图如图 4.7–1 所示。

图 4.7–1

主题：城市剪影　　**软件：**Midjourney
光圈值：f / 8.0　　**感光度：**ISO 400　　**焦距：**24mm
色彩：暖色为主　　**光影：**灯光　　**构图：**远景

提示词：City skyline at night, vibrant city lights reflecting on the water, pink and purple hues in sky, iconic buildings illuminated against dark background, wide angle view capturing skyscrapers, digital art style, high resolution, 16K, precise ISO 400 settings, shot on 24mm, f / 8.0, --ar 3:2 --v 6.0

夜晚的城市天际线，充满活力的城市灯光反射在水面上，天空是粉红色和紫色的色调，在黑暗背景下照亮的标志性建筑，捕捉摩天大楼的广角视角，数字艺术风格，高分辨率，16K，精确的 ISO 400 设置，焦距 24mm，光圈值 f / 8.0，出图比例 3∶2，版本 v 6.0

其他模板：

模板①提示词：A silhouette of an individual sitting by the river, gazing at city lights reflecting on water at night, in front her there's a bustling urban skyline, illuminated by neon signs against a dark sky, precise ISO 400 settings, shot on 35mm, f / 8.0, --ar 3:4 --v 6.0

一个人坐在河边的剪影，凝视着晚上映在水面上的城市灯光，面前是熙熙攘攘的城市天际线，霓虹灯映衬着黑暗的天空，精确的 ISO 400 设置，焦距 35mm，光圈值 f / 8.0，出图比例 3∶4，版本 v 6.0

模板②提示词：A dark silhouette of the city skyline at sunset, with buildings illuminated by city lights against an orange and blue sky, 8K, precise ISO 250 settings, shot on 50mm, f / 5.6, --ar 3:2 --v 6.0

日落时城市天际线的黑色剪影，城市灯光照亮了建筑物并映衬着橙色和蓝色的天空，8K，精确的 ISO 250 设置，焦距 50mm，光圈值 f / 5.6，出图比例 3∶2，版本 v 6.0

例：替换成模板①，如图 4.7–2 所示。

图 4.7–2

Tips

通常情况下，剪影摄影是一种通过突出主体的黑色轮廓来创造视觉冲击的摄影方式。拍摄风景剪影效果时，要注意时间、背景的选择，以及角度构图。时间通常选择黄昏时分或夜晚最佳，背景要求明亮、干净。推荐采用仰拍或平拍的形式，因为低角度更能突出主体，增加剪影效果。

4.8 哥特式建筑

哥特式建筑效果图如图 4.8–1 所示。

图 4.8–1

主题：哥特式建筑　软件：Midjourney

光圈值：f / 8.0　感光度：ISO 150　焦距：70mm

色彩：暖色为主　光影：自然光　构图：中景

提示词：A photo of the Cologne cathedral at sunset, with its towering spires and intricate details, against an orange pink sky, gothic architecture, 16K, precise ISO 150 settings, shot on 70mm, f / 8.0, --ar 2:3 --v 6.0

一张夕阳下的科隆大教堂的照片，高耸的尖顶和错综复杂的细节，映衬着橘红色的天空，哥特式建筑，16K，精确的 ISO 150 设置，焦距 70mm，光圈值 f / 8.0，出图比例 2：3，版本 v 6.0

其他模板：

模板①提示词：The medieval architecture of the Chartres cathedral, with its twin spires and intricate rose windows, stands tall against a clear blue sky, an old postcard style photograph from around year 2000, gothic architecture, precise ISO 200 settings, shot on 50mm, f / 8.0, --ar 2:3 --v 6.0

沙特尔大教堂这个中世纪建筑，有着双尖顶和复杂的玫瑰窗，矗立在清澈的蓝天下，一张 2000 年左右的明信片风格的老照片，哥特式建筑，精确的 ISO 200 设置，焦距 50mm，光圈值 f / 8.0，出图比例 2：3，版本 v 6.0

模板②提示词：Wide angle photograph of Exeter cathedral, with people sitting on grass near it and one person riding bike next to building, clear blue sky, bright daylight, green lawn, high quality photo, high resolution, gothic architecture, 8K, precise ISO 250 settings, shot on 50mm, f / 11, --ar 3:2 --v 6.0

埃克塞特大教堂的广角照片，人们坐在附近的草地上，一个人在建筑旁边骑自行车，湛蓝的天空，明亮的日光，绿色的草坪，高质量的照片，高分辨率，哥特式建筑，8K，精确的 ISO 250 设置，焦距 50mm，光圈值 f / 11，出图比例 3：2，版本 v 6.0

例：替换成模板①，如图 4.8-2 所示。

图 4.8-2

Tips

哥特式建筑是1140年左右产生于法国的欧洲建筑风格。它由罗马式建筑发展而来，为文艺复兴建筑所继承。哥特式建筑主要用于教堂，特点是尖塔高耸、尖形拱门、大窗户及花窗玻璃。在设计中利用尖肋拱顶、飞扶壁、修长的束柱，营造出轻盈修长的飞天感。新的框架结构以增加支撑顶部的力量，予以整个建筑直升线条、雄伟的外观和内部广阔空间，常结合镶着彩色玻璃的长窗。

4.9 桥类建筑

桥类建筑效果图如图4.9-1所示。

图4.9-1

主题：桥类建筑	软件：Midjourney	
光圈值：f / 5.6	感光度：ISO 200	焦距：70mm
色彩：冷色为主	光影：灯光	构图：中景

提示词：Photograph, city bridge at night, river view, bright color, bridge design, shooting, lighting, 16K, precise ISO 200 settings, shot on 70mm, f / 5.6, --ar 3:2 --v 6.0

照片，城市大桥夜景，河景，色彩鲜艳，大桥设计，拍摄，灯光，16K，精确的ISO 200设置，焦距70mm，光圈值f / 5.6，出图比例3：2，版本v 6.0

其他模板：

模板①提示词：Photography, ancient Chinese bridge design, there are streams and pebbles under the bridge, bridge design, sunshine, shooting, lighting, front view, Chinese style, precise ISO 100 settings, shot on 50mm, f / 8.0, --ar 3:2 --v 6.0

摄影，中国古代桥梁的设计，桥下有小溪和鹅卵石，桥梁设计，阳光，拍摄，灯光，正面视角，中国风，精确的ISO 100设置，焦距50mm，光圈值f / 8.0，出图比例3：2，版本v 6.0

模板②提示词：Bridge photo, the river water flows and there is some broken stone on both sides of it, the bridge surface reflects light at dusk, buildings in the distance, in the style of long exposure photography, blue sky background, wide angle lens, blue tones, natural scenery, precise ISO 100 settings, shot on 50mm, f / 11, --ar 3:2 --v 6.0

桥梁照片，流动的河水和桥两边的碎石，黄昏时桥面反射光线，远处有建筑物，长曝光摄影风格，蓝天背景，广角镜头，蓝色调，自然风光，精确的ISO 100设置，焦距50mm，光圈值f / 11，出图比例3：2，版本v 6.0

例：替换成模板①，如图4.9–2所示。

图4.9–2

Tips

在拍摄桥梁时，可以结合水面倒影，给画面增加对称的美感，通过水面倒影，桥梁得到了复制和延伸，增加了景观的层次感和时空感，从而呈现出更加美观的形象。

第5章　静物摄影：逼真的氛围感摄影

静物摄影是与人物摄影、景物摄影相对，以无生命、可人为移动或组合的物体为表现对象的摄影。多以工业或手工制成品、自然存在的无生命物体等为拍摄题材。在真实反映被摄主体固有特征的基础上，经过创意构思，并结合构图、光线、影调、色彩等摄影手段进行艺术创作，将拍摄对象表现成具有艺术美感的摄影作品。

5.1 花卉摄影

花卉摄影效果图如图 5.1–1 所示。

图 5.1–1

主题：花卉摄影	软件：Midjourney	
光圈值：f / 11	感光度：ISO 150	焦距：70mm
色彩：暖色为主	光影：自然光	构图：近景

提示词：Photography of roses in vases, with books on a table, white background, warm atmosphere, a calendar printed with the words “365 days”, soft sunlight filters through green leaves onto the flowers, enhancing colors and textures, 8K, precise ISO 150 settings, shot on 70mm, f / 11, --ar 2:3 --v 6.0

花瓶里的玫瑰，桌子上有书，白色的背景，温暖的氛围，印着“365 天”字样的日历，柔和的阳光透过绿叶洒在花朵上，增强了色彩和纹理，8K，精确的 ISO 150 设置，焦距 70mm，光圈值 f / 11，出图比例 2 : 3，版本 v 6.0

其他模板：

模板①提示词：Real photography, pink roses bloom on the fence, pink flowers hang down from above, green leaves and white walls of houses in rural areas, sunshine and a beautiful scenery, with super resolution and super details, best color grading, 4K, precise ISO 150 settings, shot on 70mm, f / 8.0, --ar 3:2 --v 6.0

真实的摄影，栅栏上绽放的粉红玫瑰，从上方垂下的粉色花朵，农村房屋的绿叶和白墙，阳光和美丽的风景，超分辨率和超细节，最佳色彩，4K，精确的 ISO 150 设置，焦距 70mm，光圈值 f / 8.0，出图比例 3 : 2，版本 v 6.0

模板②提示词：A photo of bouquet with colorful flowers, held by an outstretched hand, background is clear blue sky and white clouds, the colors include reds and pinks for roses, peonies are yellow or white in color, light pink petunias, green leaves, precise ISO 200 settings, shot on 50mm, f / 8.0, --ar 2:3 --v 6.0

一张五颜六色的花束的照片，被一只伸出的手握住，背景是清澈的蓝天和白云，玫瑰的颜色包括红色和粉色，牡丹的颜色是黄色或白色，浅粉色的牵牛花，绿色的叶子，精确的 ISO 200 设置，焦距 50mm，光圈值 f / 8.0，出图比例 2 : 3，版本 v 6.0

例：替换成模板①，如图 5.1-2 所示。

图 5.1-2

Tips

在进行花卉摄影前，需要先想清楚如何构图，因为花卉摄影更为广泛和复杂，所以拍摄前可以先从最宽的角度进行观察，在确定花卉的形态特性后，再逐步缩小范围，找到满意的画面。如果是刚开始接触摄影，可以考虑使用三分法和中心法这两种较传统的构图方法。

5.2 插花摄影

插花摄影效果图如图 5.2–1 所示。

图 5.2–1

主题：插花摄影 **软件：Midjourney**

光圈值：f / 11 **感光度：ISO 150** **焦距：24mm**

色彩：暖色和冷色相结合 **光影：自然光** **构图：远景**

提示词：Photography of a floral designer working at a table, making a bouquet with orange chrysanthemum and yellow daisies, wearing an apron, with a minimal interior design style, from a top angle shot, beige color tone, 8K, precise ISO 150 settings, shot on 24mm, f / 11, --ar 2:3 --v 6.0

一位花艺设计师在桌子前工作的照片，用橙色的菊花和黄色的雏菊制作花束，穿着围裙，极简的室内设计风格，从顶角拍摄，米色色调，8K，精确的 ISO 150 设置，焦距 24mm，光圈值 f / 11，出图比例 2∶3，版本 v 6.0

其他模板：

模板①提示词：Flower arrangement photography, flower arrangement, French style, sunshine, gentle atmosphere, clean table top, flowers and plants, high quality, 4K, precise ISO 200 settings, shot on 70mm, f / 8.0, --ar 2:3 --v 6.0

插花摄影，插花，法式风格，阳光，温和的气氛，干净的桌面，鲜花和绿植，高品质，4K，精确的 ISO 200 设置，焦距 70mm，光圈值 f / 8.0，出图比例 2：3，版本 v 6.0

模板②提示词：A modern and elegant floral arrangement, pink orchids and green pine branches in an antique white ceramic vase, against a dark red wall, the composition is symmetrical and balanced, creating contrast between softness and sharp lines, precise ISO 200 settings, shot on 70mm, f / 8.0, --ar 3:4 --v 6.0

一个现代而优雅的插花设计，粉红色的兰花和绿色的松枝插在一个古老的白色陶瓷花瓶里，靠在深红色的墙上，构图是对称和平衡的，在柔和与尖锐的线条之间形成对比，精确的 ISO 200 设置，焦距 70mm，光圈值 f / 8.0，出图比例 3：4，版本 v 6.0

例：替换成模板①，如图 5.2–2 所示。

图 5.2–2

Tips

在进行插花摄影时，为了让花卉的形态更加突出，建议选择一个相对比较简洁的背景进行拍摄，比较常见的方式就是利用白墙。但在白墙前拍摄时需要注意，如果光线不够强，画面中的白墙可能发灰，导致整体效果比较脏，这时就需要适当地提高曝光补偿，让画面尽可能明亮。除白墙外，生活化的场景会让画面更加自然，如准备一张有质感的桌布，加上一些小装饰，画面效果会更好。

5.3 美食摄影

美食摄影效果图如图 5.3-1 所示。

图 5.3-1

主题：美食摄影　　软件：Midjourney

光圈值：f / 11　　感光度：ISO 150　　焦距：100mm

色彩：暖色为主　　光影：灯光　　构图：近景

提示词：Food photo, hamburger, dark gray desktop, black background with orange light, dome light, outline light, delicious colors, photo realistic detail, front view, surreal style, realistic, high resolution, 8K, precise ISO 150 settings, shot on 100mm, f / 11, --ar 3:4 --v 5.2

食物照片，汉堡，深灰色桌面，黑色背景和橙光，圆顶灯，轮廓光，美味的色彩，逼真的细节，前视图，超现实风格，逼真，高分辨率，8K，精确的 ISO 150 设置，焦距 100mm，光圈值 f / 11，出图比例 3：4，版本 v 5.2

其他模板：

模板①提示词：Food photography, dessert with black tea, mouthwatering, the background is a fancy restaurant, real colors, comfortable light, ultra-realistic, intricate details, high quality, 4K, precise ISO 150 settings, shot on 100mm, f / 8.0, --ar 3:4 --v 5.2

美食摄影，甜点配红茶，令人垂涎，背景是一家高档餐厅，色彩真实，光线舒适，超逼真，细节复杂，高品质，4K，精确的 ISO 150 设置，焦距 100mm，光圈值 f/8.0，出图比例 3：4，版本 v5.2

模板②提示词：Food photo, many fried chicken, various parts of the chicken, dark gray desktop, black background with orange light, dome light, delicious colors, photo realistic, surreal style, high resolution, precise ISO 150 settings, shot on 80mm, f / 11, --ar 3:4 --v 5.2

美食照片，许多炸鸡，炸鸡的各个部位，深灰色的桌面，黑色的背景与橙色的光，圆顶灯，美味的色彩，照片逼真，超现实的风格，高分辨率，精确的 ISO 150 设置，焦距 80mm，光圈值 f/11，出图比例 3：4，版本 v5.2

例：替换成模板①，如图 5.3-2 所示。

图 5.3-2

Tips

通常情况下，美食摄影需要注意背景，背景越简单越好，如白色与黑色。越简单、越干净，就越能彰显食物本身的色彩。除此之外，还需要根据每种食物不同的形态和层次，选择合适的拍摄角度，比较常用的几个角度分别为鸟瞰、水平和侧面。

5.4 产品摄影

产品摄影效果图如图 5.4–1 所示。

图 5.4–1

主题：产品摄影　**软件：**Midjourney

光圈值：f / 8.0　**感光度：**ISO 150　**焦距：**85mm

色彩：暖色为主　**光影：**灯光　**构图：**近景

提示词：Lipstick, exquisite gold wire details, the scene with Chinese traditional waves, clouds and coral as the main elements, embellished with pearl material, the background is Chinese landscape, 16K, precise ISO 150 settings, shot on 85mm, f / 8.0, --ar 2:3 --v 5.2

口红，精致的金丝细节，场景以中国传统的波浪、云彩、珊瑚为主要元素，以珍珠材质点缀，背景为中国山水，16K，精确的 ISO 150 设置，焦距 85mm，光圈值 f / 8.0，出图比例 2∶3，版本 v 5.2

其他模板：

模板①提示词：Photography, aromatherapy, scented candle, champagne tone, products photo shot, candle, palm tree leaf, top angle, bright window light shadow, realistic, 16K, precise ISO 150 settings, shot on 75mm, f / 11, --ar 2:3 --v 5.2

摄影，香薰，香薰蜡烛，香槟色调，产品摄影，蜡烛，棕榈树叶，顶部视角，明亮的窗户光影，写实风，16K，精确的 ISO 150 设置，焦距 75mm，光圈值 f/11，出图比例 2：3，版本 v 5.2

模板②提示词：Product photography, a bottle of perfume is placed on the stone, with light gold and transparent glass texture, high definition, the sunshines through the window onto it, creating soft shadows that highlight its details, surrounded by some small flowers and leaves, warm tones, precise ISO 100 settings, shot on 70mm, f / 11, --ar 3:4 --niji 5

产品摄影，一瓶香水放在石头上，淡金色和透明的玻璃质感，高清晰度，阳光透过窗户照在上面，形成柔和的阴影并突出了它的细节，周围环绕着一些小花和树叶，暖色调，精确的 ISO 100 设置，焦距 70mm，光圈值 f/11，出图比例 3：4，版本 niji 5

例：替换成模板①，如图 5.4-2 所示。

图 5.4-2

Tips

在进行产品拍摄时，为了让拍出来的作品给消费者留下深刻的印象，需要有一个饱满的构图，在摆放产品时尽量遵循居中的原则。如果要将其居左或者居右摆放，那么一定要注意保持左右两侧的对称性，让整个画面看起来饱满美观，而不要给人一种失衡的感觉。

5.5 枯山水摄影

枯山水摄影效果图如图 5.5-1 所示。

图 5.5-1

主题：枯山水摄影　　软件：Midjourney

光圈值：f / 8.0　　感光度：ISO 150　　焦距：70mm

色彩：冷色为主　　光影：自然光　　构图：中景

提示词：Karesansui, a zen garden, surrounded by rocks and sand ripples, a bonsai tree, background wall has large glass windows, minimalist interior design style with natural tones, natural light, sharp focus, high resolution professional photography, 8K, precise ISO 150 settings, shot on 70mm, f / 8.0, --ar 3:2 --v 6.0

枯山水，禅宗花园，周围环绕着岩石和沙波纹，一棵盆景树，背景墙上有几扇大的玻璃窗，极简主义的室内设计风格与中性色调，自然光，焦点清晰，高分辨率的专业摄影，8K，精确的 ISO 150 设置，焦距 70mm，光圈值 f / 8.0，出图比例 3∶2，版本 v 6.0

其他模板：

模板①提示词：Photography, Karesansui, Japanese style, courtyard, zen, atrium, realistic style, real plants, 4K, precise ISO 100 settings, shot on 50mm, f / 11, --ar 4:3 --v 6.0

摄影，枯山水，日式风格，庭院，禅宗，中庭，写实风格，真实的植物，4K，精确的 ISO 100 设置，焦距 50mm，光圈值 f / 11，出图比例 4∶3，版本 v 6.0

模板②提示词：Photography, Karesansui, Japanese style, courtyard, zen, the focal point with branches, leaves on top, white sand, round window, some gray concrete wall in the background, in between there's black marble rocks on the floor, precise ISO 100 settings, shot on 35mm, f / 11, --ar 3:4 --v 6.0

摄影，枯山水，日式风格，庭院，禅宗，以老树干为焦点，顶部有树叶，白色的沙子，圆窗，背景中有一些灰色的混凝土墙，中间的地板上有黑色的大理石，精确的 ISO 100 设置，焦距 35mm，光圈值 f / 11，出图比例 3∶4，版本 v 6.0

例：替换成模板①，如图 5.5-2 所示。

图 5.5-2

Tips

枯山水是日式园林的一种，一般由细沙和碎石铺地，再加上一些叠放有致的石组所构成的缩微式园林景观，偶尔也包含苔藓、草坪或其他自然元素。枯山水中并没有水景，通常由砂石表现，而山通常用石块表现。有时也会在沙子的表面画上纹路来表现水的流动。在拍摄时，如果想让画面看起来有趣，可以勇于尝试新鲜的视角，改变我们平常的角度，从而发现景观的不同样子。

5.6 绿植摄影

绿植摄影效果图如图 5.6-1 所示。

图 5.6-1

主题：绿植摄影　　软件：Midjourney
光圈值：f / 8.0　　感光度：ISO 100　　焦距：70mm
色彩：暖色为主　　光影：自然光　　构图：中景

提示词：A balcony with various potted plants, sunlight shining through the glass door onto green leaves and flowers on wooden stands, creating an indoor scene full of natural light, warm colors, with an impressionistic rendering of light and color, 8K, precise ISO 100 settings, shot on 70mm, f / 8.0, --ar 2:3 --v 6.0

阳台上摆放着各种盆栽植物，阳光透过玻璃门洒在木架上的绿叶和花朵上，营造出充满自然光的室内场景，暖色调，用光和色彩的印象派渲染，8K，精确的 ISO 100 设置，焦距 70mm，光圈值 f / 8.0，出图比例 2∶3，版本 v 6.0

其他模板：

模板①提示词：Photography, a balcony garden filled with vibrant flowers and lush greenery, a warm atmosphere, the plants on the black metal railing add color to the scene, while sunlight filters through the leaves, a small flower pot holds orange geraniums, 4K, precise ISO 150 settings, shot on 50mm, f / 8.0, --ar 3:2 --v 6.0

摄影，一个阳台花园充满了生机勃勃的鲜花和郁郁葱葱的绿色植物，一个温馨的氛围，黑色金属栏杆上的植物为场景增添了色彩，阳光透过树叶，一个小花盆里放着橙色的天竺葵，4K，精确的 ISO 150 设置，焦距 50mm，光圈值 f / 8.0，出图比例 3∶2，版本 v 6.0

模板②提示词：Photography, green plant corner, white ceramic vase with green plants on the table, minimalist style room, overlooking a view of sea, city buildings outside the window, bright light from the left side, natural lighting, high resolution details, precise ISO 100 settings, shot on 35mm, f / 11, --ar 3:4 --v 6.0

摄影，绿植角，桌上的白色陶瓷花瓶里放着绿色的植物，极简风格的房间，可以俯瞰窗外的大海，窗外是城市建筑，左侧明亮的光线，自然采光，高分辨率细节，精确的 ISO 100 设置，焦距 35mm，光圈值 f / 11，出图比例 3∶4，版本 v 6.0

例：替换成模板①，如图 5.6-2 所示。

图 5.6-2

Tips

在拍摄植物时，掌握合适的拍摄距离是成片的关键，如果拍摄植物局部，则近景更能展现细节，这样的方法比较适合拍一些姿态优美的植物特写。如果使用中焦拍摄，则比较适合拍成丛的植物，并且选择的植物背景要干净纯粹。远距离不利于表现物体的细节，可以通过形、色来营造情绪氛围。

5.7 珠宝摄影

珠宝摄影效果图如图 5.7–1 所示。

图 5.7–1

主题：珠宝摄影　　**软件：**Midjourney

光圈值：f / 11　　**感光度：**ISO 100　　**焦距：**100mm

色彩：暖色为主　　**光影：**环境光　　**构图：**近景

提示词：Photograph, jeweled necklace next to the roses, white silk background, 8K, precise ISO 100 settings, shot on 100mm, f / 11, --ar 2:3 --niji 6

照片，玫瑰旁边的珠宝项链，白色丝绸背景，8K，精确的 ISO 100 设置，焦距 100mm，光圈值 f / 11，出图比例 2∶3，版本 niji 6

其他模板：

模板①提示词：Photography, jewelry shooting, natural warm sunlight, light colored stones, dry plants, light yellow background, with high contrast accuracy, fine gloss, centered composition, 8K, precise ISO 100 settings, shot on 100mm, f / 11, --ar 2:3 --niji 6

摄影，珠宝拍摄，自然温暖的阳光，浅色的石头，干燥的植物，淡黄色的背景，对比度和精度高，光泽度好，居中构图，8K，精确的 ISO 100 设置，焦距 100mm，光圈值 f / 11，出图比例 2∶3，版本 niji 6

模板②提示词：Photography, a ring with an emerald and sapphires on it, in the moss of a forest clearing, surrounded by wildflowers, with light green and purple tones and sparkling diamond accents, soft natural lighting of an outdoor setting, precise ISO 150 settings, shot on 80mm, f / 8.0, --ar 3:4 --niji 6

摄影，一枚镶有祖母绿和蓝宝石的戒指，在森林空地的苔藓中，周围环绕着野花，淡绿色、紫色色调和闪闪发光的钻石色调，户外的柔和自然光线，精确的 ISO 150 设置，焦距 80mm，光圈值 f / 8.0，出图比例 3∶4，版本 niji 6

例：替换成模板①，如图 5.7-2 所示。

图 5.7-2

Tips

珠宝的拍摄过程要结合其材质、做工、品质等各方面的综合素质来考虑。不同材质和不同造型的珠宝，拍摄重点也不一样，如翡翠、红蓝宝石等，着重看其净度、透度、色泽、色彩等方面；而和田玉、羊脂玉、珍珠或木质念珠等，就要看其油润度，不同情况下，拍摄方法和打光方式都不尽相同。

5.8 瓷器摄影

瓷器摄影效果图如图 5.8–1 所示。

图 5.8–1

主题：瓷器摄影　　**软件：Midjourney**

光圈值：f / 8.0　　**感光度：ISO 200**　　**焦距：70mm**

色彩：暖色为主　　**光影：环境光**　　**构图：中景**

提示词：Photograph, a Chinese blue and white porcelain teapot sits on an antique table, surrounded by flowers and tea cups, illuminated with warm light, creating a cozy atmosphere, traditional dining in China, 4K, precise ISO 200 settings, shot on 70mm, f / 8.0, --ar 2:3 --v 6.0

照片，一只中国青花瓷茶壶放在一张古色古香的桌子上，周围是花和茶杯，被温暖的灯光照亮，营造出一种舒适的氛围，中国传统餐饮，4K，精确的 ISO 200 设置，焦距 70mm，光圈值 f / 8.0，出图比例 2∶3，版本 v 6.0

其他模板：

模板①提示词：Photography, porcelain vase, a blue and white porcelain tableware set, Chinese landscape pattern decoration, background is a dark sky blue gradient, showcasing the detailed craftsmanship of each piece with gold trim on top of the glaze in the style of Chinese porcelain artists, 8K, precise ISO 100 settings, shot on 70mm, f / 11, --ar 2:3 --v 6.0

摄影，瓷器花瓶，一套青花瓷餐具，中国山水图案装饰，背景是深蓝渐变，展示了每件瓷器的详细工艺且釉上镶有中国瓷器艺术家的风格，8K，精确的 ISO 100 设置，焦距 70mm，光圈值 f / 11，出图比例 2∶3，版本 v 6.0

模板②提示词：Porcelain bowl on the table, red background wall with orange and blue patterns, white snow covering part, branch with yellow berries in foreground, depth of field effect, soft lighting, traditional style, still life photography, precise ISO 150 settings, shot on 80mm, f / 8.0, --ar 3:4 --v 6.0

餐桌上的瓷碗，红色背景墙配橙蓝图案，白色的积雪覆盖部分，前景有黄色浆果的树枝，景深效果，柔和的灯光，传统风格，静物摄影，精确的 ISO 150 设置，焦距 80mm，光圈值 f / 8.0，出图比例 3∶4，版本 v 6.0

例：替换成模板①，如图 5.8-2 所示。

图 5.8-2

Tips

拍摄瓷器时，背景要根据瓷器的颜色进行变化。例如，在拍摄青或白釉等单色瓷器时，要着重突出其光洁晶莹的质感，这时就不宜使用鲜艳的背景，否则色彩会反射到瓷器上；而在拍摄色彩丰富的瓷器时，推荐使用中灰、米黄等色调柔和的背景，以达到良好的视觉效果。

5.9 水果摄影

水果摄影效果图如图 5.9–1 所示。

图 5.9–1

主题：水果摄影　　软件：Midjourney

光圈值：f / 11　　感光度：ISO 200　　焦距：80mm

色彩：暖色为主　　光影：环境光　　构图：近景

提示词：A bowl of kiwi fruit, several whole and sliced green tender fruits on the table, surrounded in the style of gray cloth fabric with a wooden texture background, top view composition, soft lighting, high resolution image, textured, 4K, precise ISO 200 settings, shot on 80mm, f / 11, --ar 2:3 --v 6.0

一碗猕猴桃，几颗整片的绿色嫩果放在桌子上，被木纹背景的灰色布艺面料风格环绕，俯视图，柔和的灯光，高分辨率的图像，纹理，4K，精确的 ISO 200 设置，焦距 80mm，光圈值 f / 11，出图比例 2：3，版本 v 6.0

其他模板：

模板①提示词：Two fresh cherries flying through the air, surrounded by splashing water droplets, blue background, creating an illusion of weightlessness, high-quality photography, bright colors, highlight details and textures, in the style of advertising posters, 8K, precise ISO 100 settings, shot on 100mm, f / 11, --ar 2:3 --v 6.0

两个新鲜的樱桃在空中飞舞，周围是飞溅的水滴，蓝色的背景，创造了一种失重的错觉，高质量摄影，明亮的色彩，突出细节和纹理，广告海报的风格，8K，精确的ISO 100设置，焦距100mm，光圈值f/11，出图比例2：3，版本v 6.0

模板②提示词：There is an exquisite white plate on the pink table, with strawberries arranged in it, some strawberry slices placed around them, background color is light red, top view composition, precise ISO 150 settings, shot on 50mm, f / 8.0, --ar 3:4 --v 6.0

粉红色的桌子上有一个精美的白色盘子，盘子里放着草莓，周围放着草莓片，背景颜色是浅红色，俯视图，精确的ISO 150设置，焦距50mm，光圈值f / 8.0，出图比例3：4，版本v 6.0

例：替换成模板①，如图5.9-2所示。

图5.9-2

Tips

拍摄水果时，除了要拍摄的主体以外，还可以利用陪体营造整体氛围，为画面增加浪漫的氛围。例如，利用一些细节质感较强的装饰品点缀画面。但在为画面添加装饰的同时，还要注意水果与装饰品的明暗、虚实、远近关系，不要让陪体在画面中过于突出，以免产生喧宾夺主的反面效果，影响画面整体。

第6章　动物摄影：抓拍动感画面

传统的动物摄影，大多包含哺乳类、鸟类、爬虫类、两栖类等，常见的拍摄对象包括狗、猫、鸟等供玩赏的动物。动物摄影的主要目的是收集资料、供作图鉴需要的生态记录或生态摄影所需的行动记录。而对于野生动物的摄影，除了需要摄影技巧外，还需了解拍摄对象的行为习性及生态环境的相关知识。

6.1 宠物狗摄影

宠物狗摄影效果图如图 6.1–1 所示。

图 6.1–1

主题：宠物狗摄影	软件：Midjourney	
光圈值：f / 11	感光度：ISO 200	焦距：100mm
色彩：暖色为主	光影：灯光	构图：近景

提示词：Real photo, a shiba inu posing for a Christmas photoshoot, wearing a red sweater with a white and green striped scarf, a happy expression, background is a decorated Christmas tree, bright warm lighting, high quality, professional photography with sharp focus and studio lighting, some fur textures were added around the shiba inu to make it look like it was part of its natural environment, 16K, precise ISO 200 settings, shot on 100mm, f / 11, --ar 2:3 --v 6.0

真实照片，一只柴犬的圣诞照片，穿着红色毛衣配着白绿条纹围巾，脸上露出快乐的表情，背景是一棵装饰了的圣诞树，明亮温暖的灯光，高质量，带锐利焦点和工作室灯光的专业摄影，在柴犬周围添加一些皮毛纹理、使它看起来像是自然环境的一部分，16K，精确的 ISO 200 设置，焦距 100mm，光圈值 f / 11，出图比例 2∶3，版本 v 6.0

其他模板：

模板①提示词：A border collie jumps high in the air to catch a ball, with a forest background, professional photography photo, with sharp focus on the dog's face, natural light, trees are in soft shadows in the background, with blurred foreground elements, detailed fur texture of the border collie in an outdoor environment, precise ISO 150 settings, shot on 80mm, f / 11, --ar 2:3 --v 6.0

一只在空中跳得很高以接住球的边境牧羊犬，以森林为背景，专业的摄影照片，锐利的焦点对准狗的脸，自然光，背景中的树有柔和的阴影，前景元素模糊，边境牧羊犬在户外环境下的皮毛纹理，精确的 ISO 150 设置，焦距 80mm，光圈值 f / 11，出图比例 2∶3，版本 v 6.0

模板②提示词：A white Samoyed with a smiling expression stands in the background of yellow ginkgo trees, in the style of real photography, under natural light, creating warm colors, yellow leaves are falling on the ground in the natural environment, 4K, precise ISO 100 settings, shot on 85mm, f / 11, --ar 2:3 --v 6.0

一只带着微笑的白色萨摩耶犬站在黄色银杏树的背景中，真实摄影风格，在自然光下，创造出温暖的色彩，黄叶落在自然环境中的地上，4K，精确的 ISO 100 设置，焦距 85mm，光圈值 f / 11，出图比例 2∶3，版本 v 6.0

例：替换成模板①，如图 6.1-2 所示。

图 6.1-2

Tips

在拍摄宠物时，应尽量使用自然光拍摄，闪光灯不仅容易造成红眼，还容易吓到宠物，所以可以选择明亮的自然环境、户外或有窗户的室内。拍摄动物的难点在于如何让宠物保持好姿势不动，摄影师可以在它们玩耍的过程中通过喊名字或吹口哨的形式来吸引宠物的注意力，这样不仅能捕捉到它们惊奇的表情，还能捕捉到它们自然的姿态。

6.2 宠物猫摄影

宠物猫摄影效果图如图 6.2-1 所示。

图 6.2-1

主题：宠物猫摄影　**软件**：Midjourney

光圈值：f / 8.0　**感光度**：ISO 200　**焦距**：50mm

色彩：暖色为主　**光影**：自然光　**构图**：全景

提示词：A cute cat sits on the table by the window, enjoying coffee and pastries in an urban cafe, sunlight streaming through the glass, cat's soft fur contrasts beautifully against the warm tones of the room, outside is visible through the large windows, 16K, precise ISO 200 settings, shot on 50mm, f / 8.0, --ar 2:3 --v 6.0

一只可爱的猫咪坐在靠窗的桌子上，在一家城市咖啡馆里享受着咖啡和糕点，阳光透过玻璃洒进来，猫柔软的皮毛与房间的暖色调形成了美丽的对比，窗外景色清晰可见，16K，精确的 ISO 200 设置，焦距 50mm，光圈值 f / 8.0，出图比例 2：3，版本 v 6.0

其他模板：

模板①提示词：A beautiful long-haired cat standing on a rock, with green palm trees behind it, high resolution, high quality photo, with professional photography lighting, depth and texture added to enhance the scene, soft natural light, precise ISO 100 settings, shot on 70mm, f / 11, --ar 2:3 --v 6.0

一只美丽的长毛猫站在岩石上，背后是绿色的棕榈树，高分辨率，高质量的照片，专业的摄影灯光，添加景深和纹理，柔和的自然光，精确的 ISO 100 设置，焦距 70mm，光圈值 f / 11，出图比例 2∶3，版本 v 6.0

模板②提示词：Real photo, an orange cat is sitting on the side of white flowers and a blue background, high definition photography, orange fur, white flower branches in front, bright light, elegant posture, natural colors, delicate hair texture, super details, in the style of high definition photography, 4K, precise ISO 150 settings, shot on 70mm, f / 11, --ar 2:3 --v 6.0

真实的照片，一只橘猫坐在白色的花朵和蓝色的背景边，高清摄影，橙色的皮毛，前面有白色的花枝，明亮的光线，优雅的姿势，自然的颜色，细腻的毛质，超细节，高清摄影风格，4K，精确的 ISO 150 设置，焦距 70mm，光圈值 f / 11，出图比例 2∶3，版本 v 6.0

例：替换成模板①，如图 6.2-2 所示。

图 6.2-2

Tips

在拍摄宠物猫时，很多人都会犯这样一个错误：站着拍照。高角度拍摄出的照片既有距离感，又很难拍出猫的神情；低角度的拍摄不仅可以离猫更近、方便互动；平视角度利于抓拍猫的每个神态表情。

6.3 金鱼摄影

金鱼摄影效果图如图 6.3-1 所示。

图 6.3-1

主题：金鱼摄影	软件：Midjourney	
光圈值：f / 8.0	感光度：ISO 200	焦距：100mm
色彩：暖色为主	光影：灯光	构图：近景

提示词：Photo, goldfish being poured into the water tank at the bottom,motion blur and splashes, underwater bubbles rising from behind, a cinematic moment, 16K, precise ISO 200 settings, shot on 100mm, f / 8.0, --ar 2:3 --v 5.2

照片，金鱼被倒入水箱底部，动态模糊和飞溅，水下气泡从后面升起，电影般的时刻，16K，精确的 ISO 200 设置，焦距 100mm，光圈值 f / 8.0，出图比例 2：3，版本 v 5.2

其他模板：

模板①提示词：A photo shows goldfish floating in blue water, the fish is centered within the frame, it has long fins with fine scales, focusing on capturing the delicate details of its body and fin textures, precise ISO 100 settings, shot on 100mm, f / 11, --ar 2:3 --v 6.0

一张金鱼漂浮在蓝色水中的照片，鱼在框架的中心，有着长长的鳍和细鳞，专注于捕捉金鱼身体和鳍纹理的精致细节，精确的 ISO 100 设置，焦距 100mm，光圈值 f / 11，出图比例 2∶3，版本 v 6.0

模板②提示词：A golden fish with white fins and scales in the water, the background features pebbles on both sides, green plants at the bottom of the tank, water bubbles, natural light, high definition photography, delicate details, 4K, precise ISO 150 settings, shot on 70mm, f / 11, --ar 2:3 --v 6.0

一条在水中有着白色的鳍和鳞的金鱼，背景是两侧的鹅卵石，水箱底部的绿色植物，气泡，自然光，高清摄影，精致的细节，4K，精确的 ISO 150 设置，焦距 70mm，光圈值 f / 11，出图比例 2∶3，版本 v 6.0

例：替换成模板①，如图 6.3-2 所示。

图 6.3-2

Tips

在拍摄金鱼时，不同的光圈拍摄出的效果是不同的，小光圈可以清晰地拍出水里的泡泡或菱形光。但大光圈可以为画面制造出朦胧梦幻的效果，具体可以根据自己的需求而定。

6.4 狐狸摄影

狐狸摄影效果图如图 6.4–1 所示。

图 6.4–1

主题：狐狸摄影	软件：Midjourney	
光圈值：f / 11	感光度：ISO 100	焦距：50mm
色彩：暖色为主	光影：自然光	构图：全景

提示词：Photo, a red fox is lying on the ground, with leaves covering its body, tail curled up next to it, the background features green walls and brown trees, 4K, precise ISO 100 settings, shot on 50mm, f / 11, --ar 3:2 --v 6.0

照片，一只红色的狐狸躺在地上，叶子覆盖着它的身体，尾巴蜷曲在旁边，背景是绿色的墙和棕色的树，4K，精确的 ISO 100 设置，焦距 50mm，光圈值 f / 11，出图比例 3∶2，版本 v 6.0

其他模板：

模板①提示词：A cute white fox with fluffy fur sits in the snow of Alaska's wilderness, its coat shimmering under sunlight, the background is a dense forest shrouded in winter mist, creating an enchanting atmosphere, soft natural lighting, precise ISO 100 settings, shot on 70mm, f / 11, --ar 2:3 --v 6.0

一只毛茸茸的可爱的白狐坐在阿拉斯加荒野的雪地里，它的皮毛在阳光下闪闪发光，背景是笼罩在冬季的薄雾中的浓密的森林，营造出一种迷人的氛围，柔和的自然光线，精确的 ISO 100 设置，焦距 70mm，光圈值 f / 11，出图比例 2∶3，版本 v 6.0

模板②提示词：Real photo, a red fox with its head raised, looking at the camera from behind white snow in winter, with a closeup of the front view and symmetrical composition, the orange fur on his body is bright under natural light, 4K, precise ISO 150 settings, shot on 70mm, f / 11, --ar 5:3 --v 6.0

真实的照片，一只红色的狐狸抬起头，在冬天的白雪后面看着相机，正面特写和对称构图，身上的橙色皮毛在自然光下很亮，4K，精确的 ISO 150 设置，焦距 70mm，光圈值 f / 11，出图比例 5∶3，版本 v 6.0

例：替换成模板①，如图 6.4-2 所示。

图 6.4-2

Tips

在拍摄狐狸时，需要了解狐狸的基本信息，其中包括狐狸的习性、狐狸的生物学分类体系、狐狸的地理分布，并且需要长期且耐心地蹲守，学会伪装自己、隐藏自己，这样才能在不惊扰动物的情况下拍下狐狸最生动、自然的一面。

6.5 大象摄影

大象摄影效果图如图 6.5-1 所示。

图 6.5-1

主题：大象摄影　　软件：Midjourney

光圈值：f / 8.0　　感光度：ISO 400　　焦距：80mm

色彩：暖色为主　　光影：自然光　　构图：远景

提示词: A group of elephants walking across the Savannah, with acacia trees in the background, sun rays piercing through trees, creating an ethereal atmosphere, 8K, precise ISO 400 settings, shot on 80mm, f / 8.0, --ar 3:2 --v 6.0

一群大象穿过疏林草原，背景是金合欢树，阳光穿过树木，营造出一种空灵的气氛，8K，精确的 ISO 400 设置，焦距 80mm，光圈值 f / 8.0，出图比例 3∶2，版本 v 6.0

其他模板:

模板①提示词: An elephant standing by the water, photo realistic landscapes, natural light, clear skies, African Savannah, green vegetation and trees in the background, realist detail, precise ISO 250 settings, shot on 70mm, f / 8.0, --ar 2:3 --v 6.0

站在水边的大象，逼真的风景，自然光，晴朗的天空，非洲大草原，背景有绿色植被和树木，真实的细节，精确的 ISO 250 设置，焦距 70mm，光圈值 f / 8.0，出图比例 2∶3，版本 v 6.0

模板②提示词: Photo of elephants in the Savannah, with warm tones and soft lighting creating a serene atmosphere, highlight natural elements such as trees and grasslands, adding depth and texture to its background, 8K, precise ISO 400 settings, shot on 80mm, f / 8.0, --ar 3:2 --v 6.0

大象在草原上的照片，温暖的色调和柔和的灯光营造出宁静的气氛，突出树木和草地这样的自然元素，增加深度和纹理的背景，8K，精确的 ISO 400 设置，焦距 80mm，光圈值 f / 8.0，出图比例 3∶2，版本 v 6.0

例：替换成模板①，如图 6.5-2 所示。

图 6.5-2

Tips

外出拍摄大象等野生动物时，长焦镜头是必备的装备，在不打扰动物的正常生活、捕捉最自然的一面的同时，也能保障摄影师的人身安全。除此之外，野外摄影最好选择带有天窗的车，这样可以有更多的拍摄可能性，如从下往上仰拍，透过天窗拍摄停留在车顶的小动物等。

6.6 企鹅摄影

企鹅摄影效果图如图 6.6–1 所示。

图 6.6–1

主题：企鹅摄影　　**软件：Midjourney**

光圈值：f / 8.0　　**感光度：ISO 200**　　**焦距：80mm**

色彩：暖色为主　　**光影：自然光**　　**构图：全景**

提示词：A group of penguins standing in front of snow–capped mountains, the dusk sky is purple and pink, they have black eyes, white beaks and necks, real photography style, fantastic atmosphere, 8K, precise ISO 200 settings, shot on 80mm, f / 8.0, ––ar 3:2 ––v 6.0

一群企鹅站在白雪皑皑的山前，黄昏的天空是紫色和粉红色的，它们有黑色的眼睛、白色的喙和脖子，真实的摄影风格，梦幻般的氛围，8K，精确的 ISO 200 设置，焦距 80mm，光圈值 f / 8.0，出图比例 3∶2，版本 v 6.0

其他模板：

模板①提示词：A group of fluffy baby penguins, all standing on the ice, with their wings spread out, flapping them up against each other, the background is white snow, precise ISO 150 settings, shot on 100mm, f / 8.0, ––ar 3:2 ––v 6.0

一群毛茸茸的小企鹅，站在冰上，展开翅膀，互相拍打着，背景是白色的雪，精确的 ISO 150 设置，焦距 100mm，光圈值 f / 8.0，出图比例 3∶2，版本 v 6.0

模板②提示词：A penguin stands on the edge of an ice, arms outstretched to balance itself while sliding down it, the background is the blue sea with gentle waves in view, creating a spectacular scene of them gliding over snow–covered hills, 8K, precise ISO 100 settings, shot on

80mm, f / 8.0, --ar 2:3 --v 6.0

一只企鹅站在冰的边缘，伸出双臂在滑下时保持平衡，背景是蓝色的大海、海浪柔和，在白雪覆盖的山丘上滑行的场景，8K，精确的 ISO 100 设置，焦距 80mm，光圈值 f / 8.0，出图比例 2：3，版本 v 6.0

例：替换成模板①，如图 6.6–2 所示。

图 6.6–2

Tips

在南极拍摄企鹅时，要注意千万不能开闪光灯，因为这样不仅会惊吓到企鹅，还会伤害到企鹅的眼睛，所以在拍摄企鹅时，需要与它们保持特定的距离，不得触摸、环绕包围及喂食。

6.7 麋鹿摄影

麋鹿摄影效果图如图 6.7–1 所示。

图 6.7–1

主题：麋鹿摄影　　软件：Midjourney

光圈值：f / 8.0　　感光度：ISO 200　　焦距：70mm

色彩：暖色为主　　光影：自然光　　构图：中景

提示词：A majestic stag with impressive antlers, stands in the snowy landscape, bathed in the style of golden sunlight at dawn, noble presence, it roams through an expansive field covered in frosty grasses and wildflowers, focusing on its face, 16K, precise ISO 200 settings, shot on 70mm, f / 8.0, --ar 3:2 --v 6.0

一只长着令人印象深刻的鹿角的雄壮雄鹿，站在雪地上，沐浴在黎明金色的阳光中，高贵的形象，在覆盖着结霜的草和野花的广阔田野上漫步，聚焦在脸上，16K，精确的ISO 200设置，焦距70mm，光圈值f / 8.0，出图比例3：2，版本v 6.0

其他模板：

模板①提示词：Photo of a deer in the field at sunset, during the golden hour sunset, with a beautiful orange sky, tall trees, antlers towering over it, it feels ethereal and magical, precise ISO 250 settings, shot on 85mm, f / 8.0, --ar 2:3 --v 6.0

夕阳下田野里的鹿，在夕阳的黄金时刻，美丽的橘黄色天空，高大的鹿角高耸在它身上，给人一种空灵而神奇的感觉，精确的ISO 250设置，焦距85mm，光圈值f / 8.0，出图比例2：3，版本v 6.0

模板②提示词：Photo of a red deer standing in the frosty meadow at dawn, its antlers glistening with morning dew and mist rising from the grasses around it, the sun is just peeking over the horizon, 4K, precise ISO 100 settings, shot on 70mm, f / 8.0, --ar 3:2 --v 6.0

黎明时分一只鹿站在霜冻的草地上的照片，它的鹿角上闪烁着晨露，周围草地上升起了雾气，太阳刚刚从地平线上升起来，4K，精确的ISO 100设置，焦距70mm，光圈值f / 8.0，出图比例3：2，版本v 6.0

例：替换成模板①，如图6.7-2所示。

图6.7-2

Tips

在拍摄麋鹿时，我们常常会用到逆光拍摄的方法来凸显鹿角的形状。而逆光拍摄时，建议同时存储 JPEG 格式和 RAW 格式。假如因为镜头吃光，画面出现眩光或造成原片出现灰蒙蒙的情况，这时就需要后期进行色彩或锐化等处理。RAW 格式的超大宽容度会让后期的发展空间更大。

6.8 北极熊摄影

北极熊摄影效果图如图 6.8–1 所示。

图 6.8–1

主题：北极熊摄影　**软件：Midjourney**

光圈值：f / 8.0　**感光度：ISO 200**　**焦距：70mm**

色彩：冷色为主　**光影：自然光**　**构图：全景**

提示词：A polar bear standing on an ice floe, surrounded by the vast expanse of Arctic waters, the bear's white fur contrasts with the blue colors in its surroundings, it stands majestically atop the floating snow–covered iceberg, highlights the connection between wildlife and their environment, 16K, precise ISO 200 settings, shot on 70mm, f / 8.0, ––ar 3:2 ––v 6.0

一只北极熊站在一块浮冰上，周围是广阔的北极水域。这只熊的白色皮毛与周围的蓝色形成鲜明的对比，庄严地站在漂浮的冰雪覆盖的冰山上，突出了野生动物与环境之间的联系，16K，精确的 ISO 200 设置，焦距 70mm，光圈值 f / 8.0，出图比例 3 ∶ 2，版本 v 6.0

其他模板：

模板①提示词：Photo of a polar bear walking through the snow, it is a sunny day, the background features arctic ice that reflects sunlight, creating deep shadows over the snowy terrain, precise ISO 200 settings, shot on 85mm, f / 11, ––ar 10:7 ––v 6.0

北极熊在雪地里行走的照片，一个晴朗的日子，背景是反射阳光的北极冰，在积雪的地形上形成深深的阴影，精确的 ISO 200 设置，焦距 85mm，光圈值 f / 11，出图比例 10 ∶ 7，版本 v 6.0

模板②提示词：Two polar bears sleeping on an iceberg, bright color, style of real photography, 4K, precise ISO 200 settings, shot on 100mm, f / 8.0, --ar 3:2 --niji 5

两只北极熊在冰山上睡觉，明亮的色彩，真实的摄影风格，4K，精确的 ISO 200 设置，焦距 100mm，光圈值 f / 8.0，出图比例 3∶2，版本 niji 5

例：替换成模板①，如图 6.8–2 所示。

图 6.8–2

Tips

在北极进行拍摄时，光线和天气对画面的影响很大。如果想要背景呈现出通透的蓝冰，就一定要在晴空万里的天气下进行拍摄，因为阴天拍出来的照片更多是灰色的。

6.9 熊猫摄影

熊猫摄影效果图如图 6.9–1 所示。

图 6.9–1

主题：熊猫摄影　软件：Midjourney
光圈值：f / 11　感光度：ISO 100　焦距：80mm
色彩：暖色为主　光影：自然光　构图：全景

提示词：Photo, a panda sleeping in a tree, high definition, with a cute and happy expression, cute big eyes, 4K, precise ISO 100 settings, shot on 80mm, f / 11, --ar 2:3 --v 6.0

照片，一只熊猫在树上睡觉，高清分辨率，有着可爱快乐的表情，可爱的大眼睛，4K，精确的 ISO 100 设置，焦距 80mm，光圈值 f / 11，出图比例 2：3，版本 v 6.0

其他模板：

模板①提示词：A panda is sitting on tree stumps, looking down at something in front of it, the background features lush greenery and trees, precise ISO 100 settings, shot on 70mm, f / 11, --ar 3:2 --v 6.0

一只熊猫坐在树桩上，低头看着前面的东西，背景是郁郁葱葱的绿色植物和树木，精确的 ISO 100 设置，焦距 70mm，光圈值 f / 11，出图比例 3：2，版本 v 6.0

模板②提示词：A panda is eating bamboo in the green grass, chubby body, black fur on top of white fur, an adorable feeling, some willow trees around, 8K, precise ISO 150 settings, shot on 50mm, f / 11, --ar 5:3 --v 6.0

一只熊猫正在绿草地上吃竹子，胖乎乎的身体，白色的皮毛上套着黑色的皮毛，一种可爱的感觉，周围有一些柳树，8K，精确的 ISO 150 设置，焦距 50mm，光圈值 f / 11，出图比例 5：3，版本 v 6.0

例：替换成模板①，如图 6.9-2 所示。

图 6.9-2

Tips

在拍摄熊猫时，一定要耐心等待熊猫活动，把握动物的眼神和表情。例如，幼小的熊猫喜欢打斗、撒娇、爬树，这时进行拍摄，就会捕捉很多生动的画面。

6.10 火烈鸟摄影

火烈鸟摄影效果图如图 6.10-1 所示。

图 6.10-1

主题：火烈鸟摄影　　**软件：Midjourney**

光圈值：f / 8.0　　**感光度：ISO 200**　　**焦距：70mm**

色彩：暖色为主　　**光影：自然光**　　**构图：中景**

提示词：A group of flamingos playing in the water, the background is framed by green leaves, soft natural lighting highlights their vibrant colors, showcasing the beauty and grace of these birds in outdoor, high resolution, sharp focus, intricate details, hyper realistic, 16K, precise ISO 200 settings, shot on 70mm, f / 8.0, --ar 2:3 --v 5.2

一群在水中玩耍的火烈鸟，背景有绿叶的衬托，柔和的自然光突出了它们鲜艳的色彩，展示户外这些鸟的美丽和优雅，高分辨率，锐利的焦点，复杂的细节，超现实主义，16K，精确的 ISO 200 设置，焦距 70mm，光圈值 f / 8.0，出图比例 2∶3，版本 v 5.2

其他模板：

模板①提示词：A photo of many pink flamingos in the lake, cinematic, golden hour, sunset, realist detail, precise ISO 250 settings, shot on 35mm, f / 8.0, --ar 3:2 --v 6.0

一张许多火烈鸟在湖中的照片，电影感，黄金时间，日落，真实的细节，精确的 ISO 250 设置，焦距 35mm，光圈值 f / 8.0，出图比例 2∶3，版本 v 6.0

模板②提示词：A photo of flamingos in the water, with misty smoke and an ethereal atmosphere, the background features stone and greenery, a dreamlike scene, pink flamingos gracefully wading through calm waters, enchanting ambiance, a sense of tranquility and beauty, 4K, precise ISO 100 settings, shot on 70mm, f / 8.0, --ar 3:2 --v 6.0

一张火烈鸟在水中的照片，朦胧的烟雾和飘渺的气氛，背景以石头和绿色植物为主，梦幻般的场景，粉红色的火烈鸟优雅地涉水通过平静的水域，迷人的氛围，一种宁静和美的感觉，4K，精确的 ISO 100 设置，焦距 70mm，光圈值 f / 8.0，出图比例 3 : 2，版本 v 6.0

例：替换成模板①，如图 6.10-2 所示。

图 6.10-2

Tips

在拍摄鸟类生物时，我们需要配合鸟类的习性。通常情况下，野生鸟类习惯在早上及黄昏时刻觅食，在下午休息。如果想拍摄飞行、觅食等动态，需要注意拍摄时间。另外，在湿地觅食的鸟类活动会受潮汐影响，如在大潮退时会显现出大片的沼泽，鸟类就会在水面上进行觅食；相反，在大潮涨时，潮水将沼泽泥地全部覆盖，部分小型鸟类会到树上休息。

6.11 鲸鱼摄影

鲸鱼摄影效果图如图 6.11-1 所示。

图 6.11-1

主题：鲸鱼摄影	**软件：**Midjourney	
光圈值：f / 8.0	**感光度：**ISO 200	**焦距：**100mm
色彩：暖色为主	**光影：**自然光	**构图：**中景

提示词：Real photo, killer whales jumps out of the water, seagulls are flying over the sea, sunset sky, closeup shot of its mouth and fin, high definition photography, realism in the style of high definition photography, 16K, precise ISO 200 settings, shot on 100mm, f / 8.0, --ar 3:2 --v 6.0

真实的照片，虎鲸跃出水面，海鸥在海面上飞翔，日落天空，特写其嘴和鳍，高清摄影，现实主义风格的高清摄影，16K，精确的 ISO 200 设置，焦距 100mm，光圈值 f / 8.0，出图比例 3 : 2，版本 v 6.0

其他模板：

模板①提示词：Blue whales swim on the bottom of the sea, bottom of the sea, bottom creatures, goldfish and corals, a light falls, intricate details, precise ISO 250 settings, shot on 100mm, f / 8.0, --ar 2:3 --v 6.0

在海底游泳的蓝鲸，海底，海底生物，金鱼和珊瑚，一束光落下来，复杂的细节，精确的 ISO 250 设置，焦距 100mm，光圈值 f / 8.0，出图比例 2 : 3，版本 v 6.0

模板②提示词：A whale in the sea, with snowcapped mountains visible on land in background, the water is deep blue and calm as the fin glides through it, small fishing boats and buildings along shore in the distance, photo taken from behind or side of animal, focusing on its long flippers, soft pink sunset light , high resolution, realistic photo, 4K, precise ISO 150 settings, shot on 85mm, f / 11, --ar 3:2 --v 6.0

一只海里的鲸鱼，背景是陆地上的白雪皑皑的山脉，鳍滑过深蓝色、平静的海面，远处有岸边的小渔船和建筑物，从动物的背后或侧面拍摄的照片，聚焦在它的长鳍上，柔和的粉红色夕阳，高分辨率，逼真的照片，4K，精确的 ISO 150 设置，焦距 85mm，光圈值 f / 11，出图比例 3 : 2，版本 v 6.0

例：替换成模板①，如图 6.11-2 所示。

图 6.11-2

Tips

在拍摄鲸鱼时，看到跳出水面的鲸鱼要注意暂时不要接近，因为可能会被鲸鱼的大翅膀刮伤，甚至可能会被砸到，离鲸鱼尾部至少保持 5 米的安全距离；遇到母子鲸时，一般禁止下潜，因为人类下潜的动静可能会惊吓到鲸鱼宝宝，母鲸就不得不换一个地方休息。

6.12 松鼠摄影

松鼠摄影效果图如图 6.12-1 所示。

图 6.12-1

主题：松鼠摄影　**软件：**Midjourney
光圈值：f / 5.6　**感光度：**ISO 200　**焦距：**100mm
色彩：暖色为主　**光影：**自然光　**构图：**近景

提示词：A red squirrel with orange fur is sitting on top of a tree stump and eating nuts, the tail curled in the air, the background has a blurred effect to emphasize the subject, depth of field, 16K, precise ISO 200 settings, shot on 100mm, f / 5.6, --ar 2:3 --v 6.0

一只长着橙色毛的红松鼠正坐在树桩上吃坚果，尾巴在空气中打着卷，背景呈现模糊效果来强调主体，景深，16K，精确的 ISO 200 设置，焦距 100mm，光圈值 f / 5.6，出图比例 2：3，版本 v 6.0

其他模板：

模板①提示词：A red squirrel eating raspberries in the forest, on a sunny day, vibrant colors, nature photography, with natural light and a bokeh effect, beautiful background, greenery illuminated, high resolution, precise ISO 250 settings, shot on 80mm, f / 8.0, --ar 3:2 --v 6.0

一只红松鼠在森林里吃覆盆子，阳光明媚的日子里，充满活力的色彩，自然摄影，自然光和散景效果，美丽的背景，绿色植物，高分辨率，精确的 ISO 250 设置，焦距 80mm，光圈值 f / 8.0，出图比例 3：2，版本 v 6.0

模板②提示词：A red squirrel with a fluffy tail eating in a tree, on a sunny day, nature photography with green leaves in the background, 4K, precise ISO 150 settings, shot on 85mm, f / 11, --ar 3:2 --v 6.0

一只有毛茸茸尾巴的红松鼠在树上吃东西，阳光明媚的日子里，以绿叶为背景的自然摄影，4K，精确的 ISO 150 设置，焦距 85mm，光圈值 f / 11，出图比例 3：2，版本 v 6.0

例：替换成模板①，如图 6.12-2 所示。

图 6.12-2

Tips

在拍摄松鼠时，因为松鼠体型较小，运动速度较快。建议提升快门速度到至少 1/200 秒，才方便画面的捕捉。除此之外，松鼠的小体型还不方便人类的靠近，如果想要在画面中突出松鼠，可以找一个干净的背景，并配合大光圈运用浅景深效果。

第7章　人像摄影：记录最美的瞬间

人像摄影，是指借助灯光、背景、姿势等方式，以人物为主要创作对象的摄影形式。以刻画与表现被拍摄者的具体相貌和神态为主，在展现情节的基础上，着重表现被拍摄者的相貌。拍摄形式大致分为胸像、半身像、全身像。

7.1 儿童摄影

儿童摄影效果图如图 7.1–1 所示。

图 7.1–1

主题：儿童摄影	软件：Midjourney	
光圈值：f / 11	感光度：ISO 200	焦距：70mm
色彩：暖色为主	光影：灯光	构图：近景

提示词：An 8–year–old Chinese girl dressed in Chinese traditional opera costumes, photographed indoors, with bright and warm colors, half length photo, lighting effects, 8K, precise ISO 200 settings, shot on 70mm, f / 11, ––ar 3:4 ––v 5.2

一个 8 岁的穿着中国传统戏曲服装的中国女孩，室内拍摄，色彩明亮温暖，半身照片，灯光效果，8K，精确的 ISO 200 设置，焦距 70mm，光圈值 f / 11，出图比例 3：4，版本 v 5.2

其他模板：

模板①提示词：Full body photo of an 8–year–old Asian girl, with a cute and beautiful face, featuring double eyelids and big eyes, wearing yellow and standing in a sea of rapeseed flowers with a blue sky background, soft light, warm colors, in the style of high definition photography, a dreamy atmosphere, a sense of childlike innocence, 4K, precise ISO 150 settings, shot on 50mm, f / 11, --ar 2:3 --v 6.0

8岁亚洲女孩的全身照，可爱美丽的脸庞，双眼皮和大眼睛，穿着黄色站在蓝天背景的油菜花的海洋里，柔和的光线，暖色调，高清摄影的风格，梦幻的氛围，童真的感觉，4K，精确的 ISO 150 设置，焦距 50mm，光圈值 f / 11，出图比例 2∶3，版本 v 6.0

模板②提示词：Little boy dressed as spiderman standing on a tall building, with the mask in his hand, in the style of the snapshot aesthetic, dark black and red, detailed costumes, 16K, precise ISO 200 settings, shot on 35mm, f / 11, --ar 2:3 --v 6.0

小男孩装扮成蜘蛛侠站在高楼上，手里拿着面具，以快照美学的风格，深黑和红色，服装细节，16K，精确的 ISO 200 设置，焦距 35mm，光圈值 f / 11，出图比例 2∶3，版本 v 6.0

例：替换成模板①，如图 7.1–2 所示。

图 7.1–2

Tips

儿童摄影不同于普通的人像拍摄，大多数孩子天真活泼，仅靠摆拍出来的效果，很难真实地还原孩子的特点，所以推荐尝试抓拍的形式。但需要注意画面中孩子的专注点，可以给孩子一些玩具辅助拍摄，尽量不要让孩子过于在意镜头。

7.2 古风摄影

古风摄影效果图如图 7.2-1 所示。

图 7.2-1

主题：古风摄影 **软件：Midjourney**

光圈值：f / 11 **感光度：ISO 200** **焦距：70mm**

色彩：暖色和冷色相结合 **光影：自然光** **构图：中景**

提示词：Chinese beauty, wearing red and white Hanfu, with black hair in an updo style, adorned with exquisite headdress and earrings, standing on the snow-covered ground, background features falling snowflakes, creating a serene atmosphere, in the style of classical Chinese art, 8K, precise ISO 200 settings, shot on 70mm, f / 11, --ar 2:3 --v 6.0

中国美女，穿着红白相间的汉服，梳着黑色的发髻，戴着精致的头饰和耳环，站在白雪覆盖的地上，背景是飘落的雪花，营造出一种宁静的氛围，具有中国古典艺术的风格，8K，精确的 ISO 200 设置，焦距 70mm，光圈值 f / 11，出图比例 2∶3，版本 v 6.0

其他模板：

模板①提示词：Chinese girl with long hair, wearing white Hanfu and earrings, exquisite facial features, delicate skin texture, warm colors, Chinese style portrait style, light red and amber, Han Dynasty style, 4K, precise ISO 150 settings, shot on 80mm, f / 8.0, --ar 2:3 --v 6.0

留着长发的中国女孩，身穿白色汉服并戴着耳环，五官精致，肌肤细腻，暖色调，中式写真风格，淡红琥珀色，汉代风格，4K，精确的 ISO 150 设置，焦距 80mm，光圈值 f / 8.0，出图比例 2∶3，版本 v 6.0

模板②提示词：Ancient costume beautiful woman, in the style of Chinese style, wearing beige Hanfu, with white embroidered fan-shaped pattern on her chest, holding an embroidery fabric fan in hand, sitting at the eaves of traditional Chinese architecture, smiling and looking sideways at camera, with black hair curled up behind shoulders, ancient courtyard background, high definition photography, super detailed, natural light, full body portrait, 4K, precise ISO 200 settings, shot on 50mm, f / 11, --ar 2:3 --v 6.0

古装美女，中国风，身穿米黄色的汉服，胸前绣着白色的扇形图案，手里拿着绣花布扇，坐在中国传统建筑的屋檐下，微笑着侧视镜头，黑色的头发卷在肩膀后面，古代庭院背景，高清摄影，超细致，自然光，全身画像，4K，精确的 ISO 200 设置，焦距 50mm，光圈值 f / 11，出图比例 2∶3，版本 v 6.0

例：替换成模板①，如图 7.2-2 所示。

图 7.2-2

Tips

在进行古风摄影的创作时，单靠拍摄对象的造型是远远不够的，场景的选择同样重要，可以选择园林、庭院、古道城墙这些包括了中国风元素的场景。除此之外，构图也同样需要留意，可以多尝试中心、三分法、框架或者大场景的构图方式，在突出人物主体的同时为画面增添唯美、朦胧的感觉。

7.3 棚拍人像

棚拍人像效果图如图 7.3–1 所示。

图 7.3–1

主题：棚拍人像	软件：Midjourney	
光圈值：f / 11	感光度：ISO 200	焦距：50mm
色彩：暖色和冷色相结合	光影：自然光	构图：中景

提示词：Side shot, full body photo of an Asian girl with long black hair, wearing black short boots and a dark jacket sitting on top of the chair, the background is pure gray, bright makeup on her face, with high saturation colors, soft lighting, indoor studio shot, 8K, precise ISO 200 settings, shot on 50mm, f / 11, --ar 2:3 --v 6.0

侧面照，全身照，一个黑色长发的亚洲女孩，穿着黑色短靴和深色夹克，坐在椅子上，背景是纯灰色的，脸上化着明亮的妆，色彩饱和度高，光线柔和，室内摄影棚拍摄，8K，精确的 ISO 200 设置，焦距 50mm，光圈值 f / 11，出图比例 2 ：3，版本 v 6.0

其他模板：

模板①提示词：Beautiful Chinese girl, wearing a white sweater, long bangs and shoulder length hair, delicate makeup, black background, indoor studio shot, soft light, side lighting, 4K, precise ISO 150 settings, shot on 80mm, f / 8.0, --ar 2:3 --v 6.0

美丽的中国女孩，穿着白色毛衣，长刘海和齐肩长发，精致的妆容，黑色背景，室内摄影棚拍摄，光线柔和，侧面照明，4K，精确的 ISO 150 设置，焦距 80mm，光圈值 f / 8.0，出图比例 2：3，版本 v 6.0

模板②提示词：Photo of a beautiful female model, exquisite features, perfect body proportions, wearing a gray long sweater with a short skirt, soft lighting, full body shot, indoor studio shot, solid color background, in high definition photography, 4K, precise ISO 200 settings, shot on 50mm, f / 11, --ar 3:5 --v 6.0

美丽的女模特的照片，精致的五官，完美的身体比例，穿着灰色的长毛衣和短裙，光线柔和，全身照片，室内摄影棚拍摄，纯色为背景，高清照片，4K，精确的 ISO 200 设置，焦距 50mm，光圈值 f / 11，出图比例 3：5，版本 v 6.0

例：替换成模板①，如图 7.3-2 所示。

图 7.3-2

Tips

棚拍摄影的优点之一是可以根据想要的效果用灯布光。不同的人像布光法，能表现出不同效果的人物形象。基本布光包含伦勃朗光、蝴蝶光等。但同样地，布光只是辅助摄影的一种手段，应当在理解原理后灵活运用。比起布光的手段，理解并构思人物的表现手法才是关键。

7.4 街拍人像

街拍人像效果图如图 7.4–1 所示。

图 7.4–1

主题：街拍人像	软件：Midjourney	
光圈值：f / 11	感光度：ISO 150	焦距：70mm
色彩：暖色为主	光影：自然光	构图：近景

提示词：A stylish Asian woman in denim jacket, carrying coffee across the street, with buildings and cars in the background, walking down the street, sophisticated makeup, profile shot, 8K, precise ISO 150 settings, shot on 70mm, f / 11, --ar 3:2 --v 5.2

一个穿着牛仔夹克的时髦亚洲女人，端着咖啡穿过街道，背景是建筑和汽车，走在街上，精致的妆容，侧面照，8K，精确的 ISO 150 设置，焦距 70mm，光圈值 f / 11，出图比例 3 ：2，版本 v 5.2

其他模板：

模板①提示词：A stylish woman in a thick white coat, hat, full body shot, walking down the street, sophisticated makeup, ultra–realistic, intricate details, high quality, 4K, precise ISO 150 settings, shot on 35mm, f / 8.0, --ar 2:3 --v 5.2

一个穿着厚的白风衣的时尚女人，戴着帽子，全身照，走在街上，精致的妆容，超逼真，复杂的细节，高质量，4K，精确的 ISO 150 设置，焦距 35mm，光圈值 f / 8.0，出图比例 2 ：3，版本 v 5.2

模板②提示词：A stylish woman in a brown trench coat, full body shot, walking down the street, sophisticated makeup, 4K, precise ISO 100 settings, shot on 50mm, f / 11, --ar 2:3 --v 5.2

一个穿着棕色大衣的时尚女人，全身照，走在街上，精致的妆容，4K，精确的 ISO 100 设置，焦距 50mm，光圈值 f / 11，出图比例 2 ：3，版本 v 5.2

例：替换成模板①，如图 7.4-2 所示。

图 7.4-2

Tips

在进行街拍时，除了人物本身，还要结合背景。例如，将相机贴近墙面，制造纵深感和空间感。如果遇到光线很好但景色一般的位置，可以让人物遮挡住不好看的景色，留出一点靓丽的景色。

7.5 人物情绪摄影

人物情绪摄影效果图如图 7.5-1 所示。

图 7.5-1

主题：人物情绪摄影　　软件：Midjourney
光圈值：f / 8.0　　感光度：ISO 200　　焦距：70mm
色彩：暖色和冷色相结合　　光影：自然光　　构图：近景

提示词：Character emotional photography, a beautiful Chinese girl, with delicate skin, wearing white cloth and earrings, looks at the camera from side, with sunlight shining on her hair through the window, soft tones, highlights her light makeup and elegant posture, 8K, precise ISO 200 settings, shot on 70mm, f / 8.0, --ar 2:3 --v 6.0

人物情绪摄影，一个美丽的中国女孩，皮肤娇嫩，穿着白色衣服并戴着耳环，从侧面看着镜头，阳光透过窗户照在她的头发上，柔和的色调，突出了她的淡妆和优雅的姿势，8K，精确的 ISO 200 设置，焦距 70mm，光圈值 f / 8.0，出图比例 2：3，版本 v 6.0

其他模板：

模板①提示词：Character emotional photography, a young lady with long hair is looking out of the window, depressed, profile view, beautiful, elegant, emotive faces, dramatic chiaroscuro effects, facial close-up, full of atmosphere, strong emotional impact, 4K, precise ISO 150 settings, shot on 80mm, f / 8.0, --ar 5:3 --v 5.2

人物情绪摄影，一个长发少女望着窗外，忧郁，侧面照，美丽，优雅，感性的面孔，戏剧性的明暗对比效果，脸部特写，充满氛围，强烈的情感冲击，4K，精确的 ISO 150 设置，焦距 80mm，光圈值 f / 8.0，出图比例 5：3，版本 v 5.2

模板②提示词：A photo of an Asian woman with long black hair and bangs, wearing a striped sweater, leaning on the railing of her balcony while holding one hand to her lips in deep thought, the background is blurred, adding depth to the scene, she has a contemplative expression as she gazes out at the city lights, 4K, precise ISO 200 settings, shot on 50mm, f / 11, --ar 2:3 --v 6.0

一张留着黑色长发和刘海的亚洲女性的照片，穿着条纹毛衣，靠在阳台的栏杆上，一只手放在嘴唇上沉思，背景是模糊的，增加了场景的深度，她凝视着窗外的城市灯光、脸上带着沉思的表情，4K，精确的 ISO 200 设置，焦距 50mm，光圈值 f / 11，出图比例 2：3，版本 v 6.0

例：替换成模板①，如图 7.5-2 所示。

图 7.5-2

Tips

人物情绪摄影融合了被摄主体的情感表达、环境氛围以及拍摄者的表现手法等多方面因素。对于被摄主体来说，面部表情和眼神的表现最重要。为了突出情绪表情，不建议肢体上有复杂的摆姿，并且拍摄时最好不要用手部遮挡五官。在环境的选择上，不论是在室内还是室外，都尽量选择光线好的地方。拍摄时可以借助光影或光斑形成简单的明暗对比，增加画面的氛围感、趣味性。通过改变被摄主体面部的光线，还能创造出迥然不同的情绪氛围。

7.6 封面人物摄影

封面人物摄影效果图如图 7.6–1 所示。

图 7.6–1

主题：封面人物摄影　　软件：Midjourney
光圈值：f / 8.0　　感光度：ISO 150　　焦距：70mm
色彩：暖色为主　　光影：灯光　　构图：近景

提示词：Harper's Bazaar magazine cover shoot, half body photography, international, magic themes, bright colors, bold composition, 8K, precise ISO 150 settings, shot on 70mm, f / 8.0, --ar 3:4 --v 6.0

时尚芭莎杂志封面拍摄，半身摄影，国际化，奇幻的主题，鲜艳的色彩，大胆的构图，8K，精确的 ISO 150 设置，焦距 70mm，光圈值 f / 8.0，出图比例 3 : 4，版本 v 6.0

其他模板：

模板①提示词：Profile, mysterious backdrops, red and black, elegant clothing, graceful curves, soft focal points, poster art, shadowy drama, dramatic use of color, 32K, precise ISO 150 settings, shot on 35mm, f / 8.0, --ar 3:5 --v 5.2

轮廓，神秘的背景，红色和黑色，优雅的服装，优美的曲线，柔和的焦点，海报艺术，阴影戏剧，戏剧性的色彩运用，32K，精确的 ISO 150 设置，焦距 35mm，光圈值 f / 8.0，出图比例 3：5，版本 v 5.2

模板②提示词：High fashion photography in pastel pink and beige tones, a sculptural outfit made from ribbon material, the face is slightly visible, with closed eyes, white background, soft studio lighting and soft shadows, with clean sharp focus, in a hyper realistic style, 8K, precise ISO 100 settings, shot on 70mm, f / 11, --ar 2:3 --v 6.0

柔和的粉色和米色的高级时尚摄影，由丝带材料制成的雕塑服装，脸部微微可见，闭上眼睛，白色背景，柔和的工作室灯光和柔和的阴影，清晰的焦点，超现实主义的风格，8K，精确的 ISO 100 设置，焦距 70mm，光圈值 f / 11，出图比例 2：3，版本 v 6.0

例：替换成模板①，如图 7.6-2 所示。

图 7.6-2

Tips

时尚摄影师的技能之一是及时捕捉到服装或配饰的精髓。除此之外，还需要关注灯光和色彩，并安排一个平衡的构图，确保拍摄对象以能够传达服装细节的方式被照亮。如果在室内拍摄，可以利用设备进行辅助，如机滑轨，它可以提供比三脚架更多的拍摄角度。

7.7 制服照摄影

制服照摄影效果图如图 7.7-1 所示。

图 7.7-1

主题：制服照摄影　　软件：Midjourney
光圈值：f / 11　　感光度：ISO 150　　焦距：70mm
色彩：冷色为主　　光影：灯光　　构图：中景

提示词：Professional photography in the studio, uniform photos, half-body photos, a beautiful Chinese high school girl, with long black hair and bangs, black eyes, realistic photos, 8K, precise ISO 150 settings, shot on 70mm, f / 11, --ar 2:3 --v 6.0

专业摄影工作室，制服照，半身照，一个美丽的中国高中女生，黑色长发并留着刘海，黑色眼睛，逼真的照片，8K，精确的 ISO 150 设置，焦距 70mm，光圈值 f / 11，出图比例 2∶3，版本 v 6.0

其他模板：

模板①提示词：Chinese professional female, executives, office professional suit, with confident smile, crossed arms, long hair, office background, studio light, half body shot, photo by ZEISS, super details, 8K, precise ISO 100 settings, shot on 70mm, f / 8.0, --ar 3:4 --v 6.0

中国职业女性，高管，办公室职业装，自信微笑，双臂交叉，长发，办公室背景，工作室灯光，半身照，蔡司摄影，超细节，8K，精确的 ISO 100 设置，焦距 70mm，光圈值 f / 8.0，出图比例 3∶4，版本 v 6.0

模板②提示词：A beautiful high school girl wearing a white uniform, smiling, full body, professional photography, gray background, wear a white short dress, with a green plaid ribbon around her neck, blue trim on the collar, hair gently fell down to her shoulders in loose waves, a green bow tie with an emblem badge at her chest pocket, in the style of a Chinese high school girl, 16K, precise ISO 150 settings, shot on 50mm, f / 11, --ar 2:3 --v 6.0

一位漂亮的高中女生穿着白色的校服，微笑，全身照，专业摄影，灰色背景，穿着一件白色的短裙，脖子上系着一条绿色的格纹缎带，领子上镶着蓝色的镶边，头发松散地垂到肩上，胸前的口袋里系着一条绿色的领结、领结上别着一枚徽章，中国高中女生的风格，16K，精确的 ISO 150 设置，焦距 50mm，光圈值 f / 11，出图比例 2：3，版本 v 6.0

例：替换成模板①，如图 7.7-2 所示。

图 7.7-2

Tips

拍摄制服照时并不需要特别高端的摄影器材或技巧，但要注意场景的设置。选择拍摄场所时首先要确保光线充足、设备的角度正确。除此之外，为了避免产生太多背景噪声，最好选择一个干净、整洁的白墙拍摄场所。同时也要确保拍照地点反射的光线柔和而自然。

7.8 艺术照摄影

艺术照摄影效果图如图 7.8-1 所示。

图 7.8-1

主题：艺术照摄影　**软件：Midjourney**

光圈值：f / 8.0　**感光度：ISO 150**　**焦距：50mm**

色彩：暖色为主　**光影：自然光**　**构图：中景**

提示词：A cute girl wearing a black and white off-the-shoulder evening dress, with a bow on her head, long hair, holding a balloon and bunny, smiling, professional photography, high-definition details, movie atmosphere, 8K, precise ISO 150 settings, shot on 50mm, f / 8.0, --ar 2:3 --v 6.0

一个穿着黑白露肩晚礼服的可爱女孩，头上扎着蝴蝶结，留着长发，手里拿着气球和兔子，微笑，专业的摄影风格，高清细节，电影氛围，8K，精确的 ISO 150 设置，焦距 50mm，光圈值 f / 8.0，出图比例 2∶3，版本 v 6.0

其他模板：

模板①提示词：A beautiful woman in twenties, with long black hair, wearing an elegant tiara and gown, sitting at a table adorned with candles, wine bottles, sparkling glasses, a birthday cake, and party decorations, the background is a soft white gradient that transitions from light to dark, creating a dreamy atmosphere, "Happy Birthday" with gold foil above the photo, 4K, precise ISO

150 settings, shot on 35mm, f / 8.0, --ar 2:3 --v 6.0

一位20多岁的美丽女子，一头长长的黑发，头戴优雅的头饰、身穿礼服，坐在一张摆满蜡烛、酒瓶、闪闪发光的玻璃杯、生日蛋糕和派对装饰品的桌子旁，背景是柔和的白色渐变，从浅色过渡到深色，营造出梦幻般的氛围，照片上方有金箔纸写的“生日快乐”，4K，精确的ISO 150设置，焦距35mm，光圈值f / 8.0，出图比例2：3，版本v 6.0

模板②提示词：A beautiful women, red dress, sitting on chair, long hair, skin texture details, golden earrings, photo in the style of Kodak film, portrait photography, photo realistic, natural light, warm colors, grainy, hyper realistic, raw style, ultra detail, 8K, precise ISO 100 settings, shot on 50mm, f / 11, --ar 2:3 --v 6.0

一个美丽的女人，红色的衣服，坐在椅子上，长发，皮肤纹理细节，金色耳环，柯达胶片风格，人像摄影，照片写实，自然光，暖色，颗粒，超写实，原始风格，超细节，8K，精确的ISO 100设置，焦距50mm，光圈值f / 11，出图比例2：3，版本v 6.0

例：替换成模板①，如图7.8-2所示。

图7.8-2

Tips

一张好的照片最关键的就是构图，人物拍摄也需要注意取景。在取景时，相机的高度应该随着拍摄范围的扩大而下降。例如，在拍摄半身像时，相机的高度应与人物鼻子的高度相同；在拍摄全身照时，相机的高度应该在人物的腰部左右。

7.9 夜间人物摄影

夜间人物摄影效果图如图 7.9–1 所示。

图 7.9–1

主题：夜间人物摄影	软件：Midjourney	
光圈值：f / 5.6	感光度：ISO 400	焦距：70mm
色彩：暖色为主	光影：自然光	构图：中景

提示词：A beautiful Chinese woman wearing an evening dress stands on the Bund in Shanghai at night, illuminated by streetlights, she has long black hair, with bright makeup, the background features buildings of retro architecture and river views, creating a fashionable atmosphere, 8K, precise ISO 400 settings, shot on 70mm, f / 5.6, --ar 2:3 --niji 6

一位身着晚礼服的中国美女站在夜晚的上海外滩，被街灯照亮，她留着长长的黑发，化着明亮的妆，以复古建筑和河景为背景，营造时尚氛围，8K，精确的 ISO 400 设置，焦距 70mm，光圈值 f / 5.6，出图比例 2 ：3，版本 niji 6

其他模板：

模板①提示词：The high-rises in the background are illuminated by neon lights, a beautiful woman with long hair wearing a black top, looking back to the viewer on a bridge at night, a low angle shot, warm tone, light, 4K, precise ISO 250 settings, shot on 80mm, f / 8.0, --ar 2:3 --v 6.0

背景中的高楼被霓虹灯照亮，一位美丽的长发女子穿着黑色上衣，在夜晚的桥上回头看观众，低角度拍摄，暖色调，光线，4K，精确的 ISO 250 设置，焦距 80mm，光圈值 f / 8.0，出图比例 2 ∶ 3，版本 v 6.0

模板②提示词：A photo of an Asian girl wearing a black beret and leather jacket, holding pink roses in her hand, street photography, night scene, street lighting, city background, high resolution, portrait lens, natural light, soft tones, fashionable style, in the style of street photography, 8K, precise ISO 400 settings, shot on 50mm, f / 8.0, --ar 2:3 --v 6.0

一张亚洲女孩头戴黑色贝雷帽并穿着皮夹克的照片，手里拿着粉色玫瑰，街拍，夜景，街灯，城市背景，高分辨率，人像镜头，自然光，色调柔和，风格时尚，街头摄影的风格，8K，精确的 ISO 400 设置，焦距 50mm，光圈值 f / 8.0，出图比例 2 ∶ 3，版本 v 6.0

例：替换成模板①，如图 7.9–2 所示。

图 7.9–2

Tips

夜间拍摄人像时，因为环境的整体光线较弱，所以需要较高的进光量，推荐选择大光圈的定焦镜头进行拍摄，并且用大光圈进行拍摄还可以虚化背景，使夜晚的点状光源虚化出漂亮的光斑，让画面显得更加梦幻。

7.10 老人神态摄影

老人神态摄影效果图如图 7.10–1 所示。

图 7.10–1

主题：老人神态摄影　　**软件：Midjourney**

光圈值：f / 8.0　　**感光度：ISO 150**　　**焦距：70mm**

色彩：暖色为主　　**光影：自然光**　　**构图：中景**

提示词：A Chinese grandmother is smiling and selling green peppers at the street stall, wearing an apron, with white gloves on her hands, the background of a bustling city market, under sunset light, using Kodak film, with cinematic tones and intricate details, warm colors and textures, 8K, precise ISO 150 settings, shot on 70mm, f / 8.0, --ar 2:3 --v 6.0

一位中国老奶奶微笑着在街边小摊上卖青椒，穿着围裙，戴着白手套，背景是一个繁华的城市市场，在夕阳下，使用柯达胶片，具有电影般的色调和复杂的细节，温暖的颜色和纹理，8K，精确的 ISO 150 设置，焦距 70mm，光圈值 f / 8.0，出图比例 2：3，版本 v 6.0

其他模板：

模板①提示词：A photo of an elderly craftsman in his workshop, meticulously repairing and engaged with tools, surrounded by vintage materials, soft lighting creates shadows that accentuate textures and details, in the style of a vintage photograph, 4K, precise ISO 150 settings, shot on

75mm, f / 8.0, --ar 2:3 --v 6.0

一张年长工匠在他工作室里的照片，精心地修理和使用工具，周围是复古的材料，柔和的灯光创造出阴影，强调纹理和细节，具有复古照片的风格，4K，精确的 ISO 150 设置，焦距 75mm，光圈值 f / 8.0，出图比例 2：3，版本 v 6.0

模板②提示词：An old man selling vegetables on the street, wearing jackets and black pants, holding plastic bags in his hand, standing next to a car, photography, high definition details, city streets background, gray sky, urban landscape, wide angle lens, natural light, 8K, precise ISO 200 settings, shot on 50mm, f / 11, --ar 2:3 --v 6.0

一位在街上卖菜的老人，穿着夹克和黑色的裤子，手里拿着塑料袋，站在一辆车旁边，摄影照片，高清细节，城市街道背景，灰色的天空，城市景观，广角镜头，自然光，8K，精确的 ISO 200 设置，焦距 50mm，光圈值 f / 11，出图比例 2：3，版本 v 6.0

例：替换成模板①，如图 7.10-2 所示。

图 7.10-2

Tips

为老人拍照时，应该选择有情节、有人情味和趣味性的生活画面，这样才能拍摄到动人的瞬间。每位老人都有属于自己的个性特点、风格和气度，有的人热情开朗、有的人性格沉稳、有的人温文尔雅，摄影师应当学会观察，这样才能为每位老人打造出独一无二的摄影作品。

第8章 创意摄影：思维与技术、艺术的碰撞

创意摄影是指在记录物体本身的基础上，深入钻研被拍摄者的情感、思想、观念等，或者通过各种方式极富想象力地将现实生活中不存在的梦境、幻觉等景象体现在照片上。因此，创意摄影需要独特的拍摄手法、道具或后期技术上的创意来支持。

8.1 水下摄影

水下摄影效果图如图 8.1–1 所示。

图 8.1–1

主题：水下摄影　　软件：Midjourney

光圈值：f / 11　　感光度：ISO 200　　焦距：70mm

色彩：冷色为主　　光影：自然光　　构图：近景

提示词：A beautiful girl in a white dress, swimming under the sea, upper body photo, looking up at the camera, a pink flower in her hair, a beautiful light blue water surface, high quality texture, in the style of a movie poster, 16K, precise ISO 200 settings, shot on 70mm, f / 11, --ar 2:3 --v 6.0

一个穿着白色裙子的漂亮女孩，在海底游泳，上半身照片，抬头看着镜头，她的头发上有一朵粉红色的花，美丽的浅蓝色水面，高品质的质感，电影海报的风格，16K，精确的 ISO 200 设置，焦距 70mm，光圈值 f / 11，出图比例 2∶3，版本 v 6.0

其他模板：

模板①提示词：Photo of a mermaid swimming in turquoise water, fish tail visible at the bottom, long red fin and red hair, sunlight reflecting on clear waters, vibrant colors, serene atmosphere, 4K, precise ISO 150 settings, shot on 35mm, f / 8.0, --ar 3:4 --v 6.0

美人鱼在蓝绿色的水中游泳的照片，底部可以看到鱼尾，长长的红色鳍和红色的头发，阳光反射在清澈的水面上，色彩鲜艳，气氛宁静，4K，精确的 ISO 150 设置，焦距 35mm，光圈值 f / 8.0，出图比例 3∶4，版本 v 6.0

模板②提示词：A beautiful girl swims underwater, with a blue and white tail, light rays pass through the water, with bubbles, in a fantasy photography style, a dreamy atmosphere, ultra professional, with very detailed and cinematic qualities, 4K, precise ISO 200 settings, shot on 35mm, f / 11, --ar 2:3 --v 6.0

一个美丽的女孩在水下游泳，有一条蓝白相间的尾巴，光线穿过水面，气泡，奇幻的摄影风格，梦幻般的气氛，超级专业，非常细致和电影般的品质，4K，精确的 ISO 200 设置，焦距 35mm，光圈值 f / 11，出图比例 2∶3，版本 v 6.0

例：替换成模板①，如图 8.1-2 所示。

图 8.1-2

Tips

在水下摄影时，应设法避免光的散射。最简单的解决方式就是加强光照，所以可以选择阳光充足的正午时分进行拍摄，这时阳光的穿透力最好。如果有专门的水下摄影棚，建议光线用硬光，这样能削弱部分散射效果。

8.2 赛博朋克未来风摄影

赛博朋克未来风摄影效果图如图 8.2–1 所示。

图 8.2–1

主题：赛博朋克未来风摄影	软件：Midjourney	
光圈值：f / 8.0	感光度：ISO 400	焦距：70mm
色彩：冷色为主	光影：灯光	构图：近景

提示词：Real photography, from the front, a beautiful girl stands in front of the cyberpunk city background, delicate portrait, sweet smiling eyes, cute and colorful, high resolution, ice pink color, bright light, messy, soft dreamy depiction, eye–catching, cool color palette, surreal pop, 16K, precise ISO 400 settings, shot on 70mm, f / 8.0, ––ar 3:4 ––v 6.0

真实的摄影，正视图，一个美丽的女孩站在赛博朋克城市的背景前，精致的肖像，甜美微笑的眼睛，可爱多彩的，高分辨率，冰粉色，明亮的光线，凌乱，柔和的梦幻描绘，醒目，冷色调，超现实的流行，16K，精确的 ISO 400 设置，焦距 70mm，光圈值 f / 8.0，出图比例 3∶4，版本 v 6.0

其他模板：

模板①提示词：Cyberpunk, a child standing in the middle of the road, photo realistic, raining day, many tall buildings, deserted city buildings, future city, neon cold lighting, unreal engine, futuristic shops and bars, crowd, ultra wide shot, sharp focus, sharp light, 8K, precise ISO 400 settings, shot on 35mm, f / 8.0, ––ar 2:3 ––v 5.2

赛博朋克，一个站在路中间的孩子，逼真的照片，下雨天，许多高楼大厦，废弃的城市建筑，未来的城市，霓虹灯冷照明，虚幻引擎，未来的商店和酒吧，人群，超广角镜头，锐利的焦点，锐利的光线，8K，精确的 ISO 400 设置，焦距 35mm，光圈值 f / 8.0，出图比例 2：3，版本 v 5.2

模板②提示词：Cyberpunk, a female model in a black leather skirt and white jacket, standing next to neon lights, musical instruments and boxes of vinyl records, the scene is set against the backdrop of industrial architecture, with colorful lighting, she holds a guitar in one hand, with futuristic elements like holograms and a digital art style, 4K, precise ISO 200 settings, shot on 50mm, f / 11, --ar 2:3 --v 6.0

赛博朋克，一位女模特身穿黑色皮裙和白色夹克，站在霓虹灯、乐器和黑胶唱片盒旁边，场景以工业建筑为背景，灯光绚丽多彩，她一手拿着吉他，带有全息图和数字艺术风格等未来主义元素，4K，精确的 ISO 200 设置，焦距 50mm，光圈值 f / 11，出图比例 2：3，版本 v 6.0

例：替换成模板①，如图 8.2-2 所示。

图 8.2-2

Tips

赛博朋克类的摄影照片通常包括 4 种元素：主色调为蓝色、紫色、洋红色，多以城市、街道为主，画面繁杂，有明显灯光的夜景。除了需要自己打灯以外，还可以用 Photoshop 等后期工具进行调整。

8.3 婚纱主题摄影

婚纱主题摄影效果图如图 8.3-1 所示。

图 8.3-1

主题：婚纱主题摄影	软件：Midjourney	
光圈值：f / 11	感光度：ISO 200	焦距：50mm
色彩：冷色为主	光影：自然光	构图：中景

提示词：Wedding photo of a newly married couple in front of the sea, white wedding dress, white veil, luxurious atmosphere, upper body shot, long skirt, bright tone, super detailed, fantastic photography, 16K, precise ISO 200 settings, shot on 50mm, f / 11, --ar 2:3 --v 6.0

一对新婚夫妇在大海前的结婚照，白色的婚纱，白色的头纱，奢华的气氛，上半身拍摄，长裙，明亮的色调，超细节，梦幻般的摄影，16K，精确的 ISO 200 设置，焦距 50mm，光圈值 f / 11，出图比例 2∶3，版本 v 6.0

其他模板：

模板①提示词：Wedding photos, a married couple, exquisite theatrical lighting, medium shot, simple and elegant style, eye-catching, glittery, dreamy, essence of the moment, 4K, precise ISO 150 settings, shot on 50mm, f / 8.0, --ar 2:3 --v 5.2

婚纱照，一对新婚夫妇，精致的剧场灯光，中景，简洁优雅的风格，醒目，闪闪发光，梦幻，瞬间的精华，4K，精确的 ISO 150 设置，焦距 50mm，光圈值 f/8.0，出图比例 2∶3，版本 v 5.2

模板②提示词：A wedding photo of the married couple, walking on green grass, white wedding dress, white veil, the woman is holding flowers, the man wears a black suit, the man walks next to his wife, 4K, precise ISO 100 settings, shot on 35mm, f/11, --ar 2:3 --v 6.0

一对夫妇的新婚照片，走在绿色的草地上，白色的婚纱，白色的头纱，女人拿着鲜花，男人穿着黑色的西装，男人走在他的妻子旁边，4K，精确的 ISO 100 设置，焦距 35mm，光圈值 f/11，出图比例 2∶3，版本 v 6.0

例：替换成模板①，如图 8.3-2 所示。

图 8.3-2

Tips

在拍摄婚纱照时，最常用的灯光是三角光，又称 V 字形光。两盏灯会以 45° 同等距离照射主体。光线以平光为主，特点为明快、亮丽、鲜艳、温馨。在现在的婚纱摄影中，通常还会在被摄主体下方加一盏灯，与原光线构成三角形。这样的做法可以让被摄主体具有视觉冲击力，使照片更加干净、明快，更有利于后期制作。除此之外，在拍摄时，新人可以通过一些小道具，帮助自己摆姿势，改善不舒服、不自然的表情。另外，用小道具辅助拍摄，还能让画面更加丰富多彩。

8.4 太空主题摄影

太空主题摄影效果图如图 8.4-1 所示。

图 8.4-1

主题：太空主题摄影 **软件：Midjourney**

光圈值：f / 11 **感光度：ISO 150** **焦距：50mm**

色彩：冷色为主 **光影：自然光** **构图：全景**

提示词：A cinematic still of an astronaut standing on the moon, looking at the earth in space surrounded by white rocks and ice, in the style of Pixar's animated films, 4K, precise ISO 150 settings, shot on 50mm, f / 11, --ar 5:3 --v 6.0

一名宇航员站在月球上，看着被白色岩石和冰包围的太空里的地球，皮克斯动画电影风格，4K，精确的 ISO 150 设置，焦距 50mm，光圈值 f / 11，出图比例 5∶3，版本 v 6.0

其他模板：

模板①提示词：Astronaut themed photo, the boy is wearing a spacesuit, holding a helmet in his hand, sits inside a room that has the earth in the background and stars above him, a space station model on a table near him, in the style of 80s sci-fi movie posters, 4K, precise ISO 150 settings, shot on 50mm, f / 11, --ar 2:3 --v 6.0

以宇航员为主题的照片，男孩穿着宇航服，手里拿着头盔，坐在一个背景是地球、头顶是星星的房间里，他旁边的桌子上还有一个空间站模型，20 世纪 80 年代科幻电影海报的风格，4K，精确的 ISO 150 设置，焦距 50mm，光圈值 f / 11，出图比例 2∶3，版本 v 6.0

模板②提示词：A photo of woman laying on the moon, wearing white tulle dress and silver leggings, holding onto planet earth that is behind her, in space, black background, 8K, precise ISO 200 settings, shot on 35mm, f / 11, --ar 2:3 --v 6.0

一个女人躺在月球上的照片，穿着白色薄纱连衣裙和银色打底裤，抱着她身后的地球，在太空中，黑色背景，8K，精确的 ISO 200 设置，焦距 35mm，光圈值 f / 11，出图比例 2∶3，版本 v 6.0

例：替换成模板①，如图 8.4–2 所示。

图 8.4–2

Tips

创意摄影需要用到不同的摄影技巧，从而创造出震撼人心的视觉效果。但创新的精神不仅体现在拍摄过程中，还包括后期处理。通过后期调色、合成等方式，可以进一步强化影像的表达效果，创造出更抽象或更奇妙的作品。

8.5 童话风格摄影

童话风格摄影效果图如图 8.5–1 所示。

图 8.5–1

主题：童话风格摄影　　软件：Midjourney
光圈值：f / 11　　感光度：ISO 150　　焦距：50mm
色彩：冷色为主　　光影：自然光　　构图：中景

提示词：Real photo, an 18-year-old girl, blond hair, Alice in Wonderland, she is wearing a light blue dress, holding two little white rabbits, surrounded by flowers and mushrooms, colorful and detailed, 4K, precise ISO 150 settings, shot on 50mm, f / 11, --ar 3:4 --v 6.0

真实照片，一个18岁的女孩，金发，爱丽丝梦游仙境，她穿着一条浅蓝色的连衣裙，抱着两只小白兔，周围是鲜花和蘑菇，色彩鲜艳，细节丰富，4K，精确的ISO 150设置，焦距50mm，光圈值f/ 11，出图比例3∶4，版本v 6.0

其他模板：

模板①提示词：Little Red Riding Hood cosplay, girl picking apples from the basket in front of her, with long blonde hair and red cloak, holding an apple to her mouth, professional photography, ultra realistic, detailed, professional color grading, soft shadows, clean sharp focus, 4K, precise ISO 150 settings, shot on 70mm, f / 11, --ar 2:3 --v 6.0

小红帽的角色扮演，女孩从面前的篮子里拿苹果，金黄色的长发，红色的斗篷，拿着苹果放到嘴里，专业的摄影，超逼真，细致，专业的色彩分级，柔和的阴影，干净的锐利焦点，4K，精确的ISO 150设置，焦距70mm，光圈值f/ 11，出图比例2∶3，版本v 6.0

模板②提示词：Real photo, yellow dress, Beauty and the Beast, hyper-realistic, realistic shadow and light, high detail, 8K, precise ISO 200 settings, shot on 50mm, f / 11, --ar 2:3 --v 6.0

真实照片，黄色连衣裙，美女与野兽，超现实，逼真的光影，高细节，8K，精确的ISO 200设置，焦距50mm，光圈值f/ 11，出图比例2∶3，版本v 6.0

例：替换成模板①，如图8.5-2所示。

图8.5-2

Tips

在进行童话风格的创意摄影时，首先应该确认拍摄的主题，主题可以借鉴一些知名的童话故事。然后在故事的基础上确定拍摄场景、拍摄时间和拍摄道具。要确定方向，才能拍出想要的主题。

8.6 精灵风格摄影

精灵风格摄影效果图如图 8.6–1 所示。

图 8.6–1

主题：精灵风格摄影	软件：Midjourney	
光圈值：f / 11	感光度：ISO 200	焦距：70mm
色彩：冷色为主	光影：自然光	构图：中景

提示词：Photo of a girl in white dress, pale skin, long hair, blonde bangs, sparkling silver headband, surrounded by green bushes, fantasy style, sunlight, glittery gold lace, anime aesthetic, 4K, precise ISO 200 settings, shot on 70mm, f / 11, --ar 3:2 --v 6.0

一个穿着白色裙子的女孩的照片，苍白的皮肤，长发，金色的刘海，闪闪发光的银色发带，被绿色的灌木丛包围，梦幻般的风格，阳光下，发光的金色蕾丝，动漫美学，4K，精确的 ISO 200 设置，焦距 70mm，光圈值 f / 11，出图比例 3∶2，版本 v 6.0

其他模板：

模板①提示词：A photo of an Asian woman with long dark hair, wearing white wings, silver jewelry on her hands, wearing a short dress made out of lace that has feathers on it, sitting on the moonlight at night, fluffy clouds all around her, eyes glow golden yellow, 4K, precise ISO 150 settings, shot on 50mm, f / 8.0, --ar 2:3 --v 6.0

一张深色长发的亚洲女性的照片，戴着白色的翅膀，手上戴着银色的首饰，穿着带羽毛的蕾丝短裙，坐在夜晚的月光下，周围是蓬松的云朵，眼睛闪烁着金黄色的光芒，4K，

精确的 ISO 150 设置，焦距 50mm，光圈值 f / 8.0，出图比例 2：3，版本 v 6.0

模板②提示词：Chinese girl with long hair, wear wings on the back, in a green tulle skirt, red lips, white transparent gauze flying around her, exquisite facial features, perfect body proportions, a grassy field background, soft light, a charming posture, exquisite makeup, high-definition photography, 16K, precise ISO 100 settings, shot on 50mm, f / 11, --ar 2:3 --v 6.0

一头长发的中国女孩，背后戴着翅膀，穿着绿色薄纱裙，红唇，白色透明的纱布在她周围飞舞，精致的五官，完美的身材比例，草地背景，柔和的光线，迷人的姿势，精致的妆容，高清的摄影，16K，精确的 ISO 100 设置，焦距 50mm，光圈值 f / 11，出图比例 2：3，版本 v 6.0

例：替换成模板①，如图 8.6-2 所示。

图 8.6-2

Tips

在拍摄精灵风格的图像时，可以看出这类图像需要柔和的光影，从而制造出梦幻感。所以在拍摄时，可以多运用柔光，柔光是一种漫射光，可避免在主体上投射出棱角分明的阴影，而是逐渐而均匀的光影。这样就有利于赋予主画面柔和或浪漫的感觉。

8.7 龙年生肖主题摄影

龙年生肖主题摄影效果图如图 8.7–1 所示。

图 8.7–1

主题：龙年生肖主题摄影	软件：Midjourney	
光圈值：f / 11	感光度：ISO 150	焦距：70mm
色彩：暖色和冷色相结合	光影：灯光	构图：中景

提示词：Surrealistic photography, close–up of a beautiful girl dressed in a gorgeous Hanfu with red and a big dragon, surrounded by a giant dragon covered in ice and snow, made of jade and glass material, flowing silk, dreamy scene, magnificent scale, movie lighting effects, 4K, precise ISO 150 settings, shot on 70mm, f / 11, ––ar 2:3 ––v 6.0

超现实主义的摄影，一个美丽的女孩穿着华丽的汉服和一条大龙的特写，被一条被冰雪覆盖的巨大的龙包围，由玉石和玻璃材料制成，流动的丝绸，梦幻般的场景，宏伟的规模，电影灯光效果，4K，精确的 ISO 150 设置，焦距 70mm，光圈值 f / 11，出图比例 2∶3，版本 v 6.0

其他模板：

模板①提示词：Surrealistic children's photography, close–ups of a handsome boy standing in front of thick clouds and mist surrounded by a white dragon, in white Hanfu, with short hair, with captivating eyes and delicate features, 4K, precise ISO 150 settings, shot on 70mm, f / 11, ––ar 2:3 ––v 6.0

超现实主义的儿童摄影，一个帅气的男孩站在厚厚的云雾前，周围是一头白色的龙，穿着白色的汉服，头发短短的，眼睛迷人，五官精致，4K，精确的 ISO 150 设置，焦距 70mm，光圈值 f / 11，出图比例 2∶3，版本 v 6.0

模板②提示词：A beautiful Chinese woman, wearing an exquisite headdress and earrings on her head, with the dragon behind her as the background, red color scheme, exquisite details, symmetrical composition, movie lighting effects, soft light, portrait photography, exquisite facial features, gorgeous colors, mysterious atmosphere, 8K, precise ISO 200 settings, shot on 70mm, f / 11, --ar 2:3 --v 6.0

一位美丽的中国女子，头上戴着精美的头饰和耳环，背景是身后的龙，红色的配色，细节精致，构图对称，电影灯光效果，柔和的光线，人像摄影，长焦镜头，精致的五官，绚丽的色彩，神秘的气氛，8K，精确的 ISO 200 设置，焦距 70mm，光圈值 f / 11，出图比例 2∶3，版本 v 6.0

例：替换成模板①，如图 8.7–2 所示。

图 8.7–2

Tips

在进行生肖主题的创意摄影时，因为生肖元素和中国传统文化有关，所以还可以在画面中加入适量的中国风元素，以达到画面的平衡和自然。

8.8 花海人像摄影

花海人像摄影效果图如图 8.8–1 所示。

图 8.8–1

主题：花海人像摄影	软件：Midjourney	
光圈值：f / 11	感光度：ISO 200	焦距：70mm
色彩：暖色为主	光影：自然光	构图：中景

提示词：A Chinese girl with long hair, wearing an off–the–shoulder red dress and hat, sitting in the flower sea, soft tones, delicate skin texture, natural light, shallow depth of field, high saturation colors, elegant movements, 16K, precise ISO 200 settings, shot on 70mm, f / 11, --ar 2:3 --v 6.0

一个长发的中国女孩，穿着露肩红裙子并戴着帽子，坐在花海中，柔和的色调，细腻的皮肤质感，自然光线，浅景深，高饱和度的色彩，优雅的动作，16K，精确的 ISO 200 设置，焦距 70mm，光圈值 f / 11，出图比例 2∶3，版本 v 6.0

其他模板：

模板①提示词：A beautiful girl, with long hair, in white dress, sitting on the grass covered in violet flower petals, smiling with a smiling face, a purple field background, with natural lighting, soft tones and shadows, high details and best quality, 4K, precise ISO 150 settings, shot on 70mm, f / 8.0, --ar 2:3 --v 6.0

一个漂亮的女孩，长发飘飘，穿着白色连衣裙，坐在铺满紫罗兰花瓣的草地上，面带微笑，紫色田野背景，光线自然，色调和阴影柔和，细节高、质量好，4K，精确的 ISO 150 设置，焦距 70mm，光圈值 f / 8.0，出图比例 2∶3，版本 v 6.0

模板②提示词：Portrait photography of a Chinese girl holding flowers in the park, she is wearing a light blue long-sleeved top and white skirt, with a happy smile, full-body shot, with a spring forest background, pink and purple wildflowers are blooming, the natural lighting creates soft tones and a warm atmosphere, 16K, precise ISO 100 settings, shot on 50mm, f / 11, --ar 2:3 --v 6.0

一个中国女孩在公园里拿着花的人像摄影，她穿着浅蓝色的长袖上衣和白色的裙子，脸上带着幸福的微笑，全身照，以春天的森林为背景，粉红色和紫色的野花盛开，自然光营造出柔和的色调与温暖的氛围，16K，精确的 ISO 100 设置，焦距 50mm，光圈值 f / 11，出图比例 2∶3，版本 v 6.0

例：替换成模板①，如图 8.8-2 所示。

图 8.8-2

Tips

拍花是一个相对不挑天气的主题，通常情况下，推荐在春天拍摄，春天的光线相对柔和，阳光没有那么强烈刺眼，最佳拍摄时间在下午 3 点与早上 10 点前后，这段时间的光线是最柔和的。如果光线太强，会有很强烈的明暗阴影，导致人脸上的颗粒感特别明显。

第9章　不同艺术风格：向大师看齐

艺术风格是指摄影师在前期拍摄或后期处理中展现出的属于自己的独特个性和风格。个人风格的形成不仅靠技术的运用，还包含了每个摄影师不同的艺术表现、思维方式、审美观点，以及受社会文化背景的影响。

9.1 纪实主义

纪实主义效果图如图 9.1–1 所示。

图 9.1–1

主题：纪实主义	**软件：Midjourney**	
光圈值：f / 11	**感光度：ISO 100**	**焦距：70mm**
色彩：冷色为主	**光影：自然光**	**构图：中景**

提示词：A Chinese old man is sitting on the bench, and an elderly person wearing gloves sits in front of him with his head down, outside a store entrance next to the road in black and white photos, 16K, precise ISO 100 settings, shot on 70mm, f / 11, --ar 3:2 --v 6.0

一位中国老人坐在长凳上，一位戴着手套的老人低着头坐在他前面，一家路边的商店门口的黑白照片，16K，精确的 ISO 100 设置，焦距 70mm，光圈值 f / 11，出图比例 3：2，版本 v 6.0

其他模板：

模板①提示词：A country man was walking on a road covered with railroad tracks, the background shows houses in the distance, in a black and white style, movie stills, documentary footage in a retro style, 4K, precise ISO 150 settings, shot on 50mm, f / 8.0, --ar 2:3 --v 6.0

一个乡下的男人走在布着铁轨的路上，背景是远处的房屋，黑白风格，电影剧照，复古风格的纪录片镜头，4K，精确的 ISO 150 设置，焦距 50mm，光圈值 f / 8.0，出图比例 2 : 3，版本 v 6.0

模板②提示词：A little girl is writing at her desk, with black hair and big eyes, the camera focuses on her hand holding a pencil, in the class, in a black and white style, 4K, precise ISO 200 settings, shot on 50mm, f / 11, --ar 2:3 --v 6.0

一个小女孩正在她的课桌上写字，有黑头发和大眼睛，镜头聚焦在她握着铅笔的手上，在教室里，黑白风格，4K，精确的 ISO 200 设置，焦距 50mm，光圈值 f / 11，出图比例 2 : 3，版本 v 6.0

例：替换成模板①，如图 9.1-2 所示。

图 9.1-2

Tips

纪实主义摄影风格是以真实记录生活现实为主要目的，力求抓住生活的片段、社会风俗、人物表情和故事等各个方面，如实反映我们所看到的场景，强调视角的客观性，有记录和保存历史的价值。

9.2 印象派

印象派效果图如图 9.2–1 所示。

图 9.2–1

主题：印象派　　软件：Midjourney
光圈值：f / 8.0　　感光度：ISO 150　　焦距：35mm
色彩：冷色为主　　光影：自然光　　构图：全景

提示词：A photo of an old forest at night, cinematic still, foggy and mysterious, the sun is shining through the trees casting long shadows on the grassy ground, someone walking away, wearing black, hyper realistic, impressionist photography, 16K, precise ISO 150 settings, shot on 35mm, f / 8.0, --ar 3:2 --v 6.0

一张夜晚的古老森林的照片，电影剧照，雾蒙蒙而神秘，阳光透过树林并在草地上投下长长的影子，有人从前面走了出来，穿着黑衣服，超现实，印象派摄影，16K，精确的 ISO 150 设置，焦距 35mm，光圈值 f / 8.0，出图比例 3∶2，版本 v 6.0

其他模板：

模板①提示词：A farmer in the rice fields at sunset, wearing a traditional conical hat, holding a stick, walking through water with reeds on both sides, with distant mountains visible under an orange sky, the scene captures their silhouette against the backdrop of nature's beauty, impressionist photography, 8K, precise ISO 400 settings, shot on 50mm, f / 11, --ar 3:2 --v 6.0

一位农民在夕阳下的稻田里，戴着传统的圆锥形帽子，拿着一根棍子，在两边长满芦苇的水中行走，橙色的天空下可以看到远处的群山，捕捉到了他们在大自然美景背景下的剪影，印象派摄影，8K，精确的 ISO 400 设置，焦距 50mm，光圈值 f / 11，出图比例 3∶2，版本 v 6.0

模板②提示词：Impressionist photography, a serene landscape of an ancient Chinese lake, reflecting the misty mountains in soft hues of gray and blue, a small pavilion stands on one side, with lush green trees around it, the water reflects the sky's early morning light, creating gentle ripples that mirror distant peaks, 8K, precise ISO 200 settings, shot on 35mm, f / 11, --ar 3:4 --v 6.0

印象派摄影，中国古代湖泊的宁静景观，在灰色和蓝色的柔和色调中反映出薄雾缭绕的山脉，旁边有个小亭子，四周绿树成荫，水面反射着天空的晨光，形成轻柔的涟漪并映照出

远处的山峰，8K，精确的 ISO 200 设置，焦距 35mm，光圈值 f/11，出图比例 3：4，版本 v 6.0

例：替换成模板①，如图 9.2-2 所示。

图 9.2-2

Tips

印象派摄影的特点是色调沉郁、影纹粗糙，富有装饰性但缺乏空间感。一眼看上去似乎完全丧失了摄影艺术自身的特点，画面更像绘画作品，所以有人将其称为“仿画派”。

9.3 简约风

简约风效果图如图 9.3-1 所示。

图 9.3-1

主题：简约风　　软件：Midjourney

光圈值：f / 8.0　　感光度：ISO 150　　焦距：80mm

色彩：冷色为主　　光影：自然光　　构图：近景

提示词：A Chinese beauty sitting at the table, wearing white, holding flowers in her hand, bright eyes, the background is clean and simple, featuring light tones of beige or off-white color scheme, exquisite makeup, showcasing an elegant posture, warm atmosphere, with high resolution, 16K, precise ISO 150 settings, shot on 80mm, f / 8.0, --ar 2:3 --v 6.0

一位中国美女坐在桌旁，身穿白衣，手里拿着花，炯炯有神的眼睛，背景干净简单，浅色调的米色或灰白色的配色，精致的妆，展现出优雅的姿态，温暖的氛围，高分辨率，16K，精确的 ISO 150 设置，焦距 80mm，光圈值 f / 8.0，出图比例 2∶3，版本 v 6.0

其他模板：

模板①提示词：A Chinese beauty wearing white long sleeves, sweater with large fur material, hair in loose waves, sat on the bed with her legs crossed behind her, smiling, the background is pure white, creating an atmosphere of soft tones, smooth skin, delicate features, 8K, precise ISO 150 settings, shot on 70mm, f / 8.0, --ar 2:3 --v 6.0

一位穿着白色长袖毛茸茸毛衣的中国美女，头发蓬松，盘腿坐在床上，微笑，背景为纯白色，营造出柔和的色调氛围，光滑的皮肤，精致的五官，8K，精确的 ISO 150 设置，焦距 70mm，光圈值 f / 8.0，出图比例 2∶3，版本 v 6.0

模板②提示词：A young woman in early twenties, wearing an elegant black long-sleeved top and jeans, classic portrait photography, 8K, precise ISO 200 settings, shot on 50mm, f / 11, --ar 3:4 --v 6.0

一位二十出头的年轻女性，穿着优雅的黑色长袖上衣和牛仔裤，经典肖像摄影，8K，精确的 ISO 200 设置，焦距 50mm，光圈值 f / 11，出图比例 3∶4，版本 v 6.0

例：替换成模板①，如图 9.3-2 所示。

图 9.3-2

Tips

在进行简约风照片的拍摄时，前期可以尽可能多地尝试多角度拍摄；多观察生活中的细节，发现简约之美；除全景外，还可以多靠近主体，调大光圈，虚化背景。后期方面，可以尝试剪裁画面、进行二次构图。

9.4 复古风

复古风效果图如图 9.4–1 所示。

图 9.4–1

主题：复古风	软件：Midjourney	
光圈值：f / 8.0	感光度：ISO 150	焦距：80mm
色彩：暖色为主	光影：自然光	构图：近景

提示词：A color photo of an elegant woman, in the style from the movie Roman Holiday, with pink lips, light blue eyes, wearing white dress shirt with ruffled collar, standing on street, hair up in clean bun with small bow , background is busy outdoor piazza with people walking by, 16K, precise ISO 150 settings, shot on 80mm, f / 8.0, --ar 2:3 --v 6.0

一位优雅的女人的彩色照片，电影《罗马假日》风格，粉红色的嘴唇，浅蓝色的眼睛，穿着白色的礼服衬衫与褶边领，站在街上，干净的发髻与小蝴蝶结，背景是繁忙的户外广场与路过的人们，16K，精确的 ISO 150 设置，焦距 80mm，光圈值 f / 8.0，出图比例 2：3，版本 v 6.0

其他模板：

模板①提示词：A beautiful woman sitting in front of a mirror, wearing an Hepburn style skirt,

1950s, knee shot, contrasting light and dark tones, sharp focus, cinematic mood, black and white photo, vintage effect, photo by Kodak, classic style, 8K, precise ISO 150 settings, shot on 70mm, f / 8.0, --ar 2:3 --v 5.2

一位美丽的女人坐在镜子前，穿着赫本风格的裙子，20 世纪 50 年代，膝盖以上摄影，明暗色调对比，锐利的焦点，电影般的氛围，黑白照片，复古效果，柯达照片，经典风格，8K，精确的 ISO 150 设置，焦距 70mm，光圈值 f / 8.0，出图比例 2∶3，版本 v 5.2

模板②提示词：Photo, a beautiful girl with black long hair, curly hair, big bright eyes, retro suspender, vintage style, low saturation colors, mottled texture, the yellowed texture, front view, background is a wall composed by newspapers, 8K, precise ISO 200 settings, shot on 50mm, f / 11, --ar 3:4 --v 6.0

照片，一位有黑色长发的漂亮女孩，卷发，大而明亮的眼睛，复古的背带裤，复古的风格，低饱和度的颜色，斑驳的纹理，泛黄的纹理，正面视图，背景是由报纸组成的墙，8K，精确的 ISO 200 设置，焦距 50mm，光圈值 f / 11，出图比例 3∶4，版本 v 6.0

例：替换成模板①，如图 9.4-2 所示。

图 9.4-2

Tips

在拍摄完复古风照片后，通常需要进行后期调色，我们要时刻注意观察画面，适度调整色调。在完成对画面明暗和光影的调整后，如果想要呈现复古风的感觉，就需要对照片的色温进行修饰。可以根据情况适当地增加色温让画面变暖、变黄，再增加一些自然饱和度让色彩更加浓郁，定好基本的调子。

9.5 富士胶卷风

富士胶卷风效果图如图 9.5-1 所示。

图 9.5-1

主题：富士胶卷风　**软件**：Midjourney

光圈值：f / 8.0　**感光度**：ISO 250　**焦距**：50mm

色彩：暖色为主　**光影**：自然光　**构图**：全景

提示词：An 18-year-old Chinese girl in a tulle dress, a flock of seagulls circled around her against the setting sun, gliding gently across the sky, by the sea, the breeze blows her hair, the sun warms her face, the girl among the clouds and the seagulls, Fuji camera effect, 16K, precise ISO 250 settings, shot on 50mm, f / 8.0, --ar 2:3 --v 6.0

一位 18 岁的中国女孩穿着薄纱连衣裙，一群海鸥在夕阳的映衬下绕着她盘旋，轻轻地划过天空，在海边，微风吹拂着她的头发，阳光温暖着她的脸，女孩在云和海鸥之间，富士相机效果，16K，精确的 ISO 250 设置，焦距 50mm，光圈值 f / 8.0，出图比例 2∶3，版本 v 6.0

其他模板：

模板①提示词：A close-up of the face and neck, a beautiful woman with black hair is standing in heavy rain at night, under street lights, tilted her head to look up at the sky, the water droplets reflecting the soft glow around her, background blurs into a warm yellow color, her expression reflects tranquility and serenity, Fuji camera effect, precise ISO 400 settings, shot on 80mm, f / 11,

--ar 2:3 --v 6.0

一张脸部和颈部的特写，一位黑发美女站在夜晚的大雨中，站在路灯下，抬着头仰望天空，水滴在她周围反射出柔和的光芒，背景模糊成温暖的黄色，她的表情平静和安宁，富士相机效果，精确的 ISO 400 设置，焦距 80mm，光圈值 f / 11，出图比例 2∶3，版本 v 6.0

模板②提示词：A beautiful woman with long hair, wearing an amber knit sweater and black jeans, is taking out food from the refrigerator in her home kitchen, she has both hands holding some snacks in paper bags, sunlight streaming through the window onto her face, Fuji camera effect, precise ISO 150 settings, shot on 50mm, f / 11, --ar 3:4 --v 6.0

一位漂亮的长发女人，穿着琥珀色针织毛衣和黑色牛仔裤，正从自家厨房的冰箱里取出食物，她双手拿着纸袋里的零食，阳光透过窗户照在她的脸上，富士相机效果，精确的 ISO 150 设置，焦距 50mm，光圈值 f / 11，出图比例 3∶4，版本 v 6.0

例：替换成模板①，如图 9.5-2 所示。

图 9.5-2

Tips

在使用胶片进行拍照时，要注意避光保存未曝光的与曝光未冲洗的胶片，在夏天还要注意避免高温储存。潮湿的环境也会对胶片产生影响，因为明胶遇水会变软膨胀，很容易从胶片上脱落。感光乳剂也容易受潮而变质。

9.6 特色民族风格

特色民族风格效果图如图 9.6–1 所示。

图 9.6–1

主题：特色民族风格	软件：Midjourney	
光圈值：f / 8.0	感光度：ISO 150	焦距：50mm
色彩：暖色为主	光影：自然光	构图：中景

提示词：A beautiful girl dressed in the traditional costumes of Yunnan's ethnic Dai people, with fur and colorful patterns on her, stood gracefully playing with white sheep under the sunshine at the grassland, she has long hair tied into an intricate braid adorned with vibrant ribbons, backdrop features majestic mountains, 16K, precise ISO 150 settings, shot on 50mm, f / 8.0, --ar 2:3 --v 6.0

一个穿着云南傣族传统服装的美丽女孩，身上有毛皮和五颜六色的图案，优雅地在阳光下的草原上和白羊玩耍，她的长发编成了一个复杂的辫子、上面装饰着鲜艳的丝带，背景以雄伟的山脉为特色，16K，精确的 ISO 150 设置，焦距 50mm，光圈值 f / 8.0，出图比例 2∶3，版本 v 6.0

其他模板：

模板①提示词：Beautiful Chinese girl in an ethnic style skirt, headband in her hair, earrings on her ears, wearing colorful, red lips, long black wavy hair, gold accessories around her neck and forehead, night sky background with city lights, 8K, precise ISO 200 settings, shot on 70mm, f / 8.0, --ar 2:3 --v 6.0

美丽的中国女孩穿着民族风格的裙子，头上扎着发带，耳朵上戴着耳环，穿得五颜六色，红唇，长长的黑色卷发，脖子和额头上戴着金色的饰品，夜空背景是城市的灯光，8K，精确的 ISO 200 设置，焦距 70mm，光圈值 f / 8.0，出图比例 2：3，版本 v 6.0

模板②提示词：A beautiful Chinese woman wearing white and red traditional Miao, with silver tassels on her head hat, old stone walls, real photography effects, warm tones, natural light, soft shadows, precise ISO 150 settings, shot on 50mm, f / 11, --ar 3:2 --v 6.0

一位美丽的中国女人穿着红白相间的传统苗族服饰，头上戴着银色流苏的帽子，古老的石墙，真实摄影效果，温暖的色调，自然光，柔和的阴影，精确的 ISO 150 设置，焦距 50mm，光圈值 f / 11，出图比例 3：2，版本 v 6.0

例：替换成模板①，如图 9.6–2 所示。

图 9.6–2

Tips

在进行民族服饰的人像摄影时，可以事先准备一些符合服饰风格的小道具，这样在拍摄时才不会出现傻站着不知道手脚往哪里放的情景。同时，对服饰和道具的文化还要足够了解，这样才能给自己定一个风格基调，让自己沉浸到那个角色里，从而拍出适合自己风格的摄影作品。

9.7 青春校园风格

青春校园风格效果图如图 9.7–1 所示。

图 9.7–1

主题：青春校园风格	软件：Midjourney	
光圈值：f / 8.0	感光度：ISO 150	焦距：70mm
色彩：暖色为主	光影：自然光	构图：中景

提示词：A Chinese high school girl in white and blue sportswear sits on the red track, the background is grassy ground with sports equipment and stands, her hair was tied back, wearing sneakers, delicate facial features, 16K, precise ISO 150 settings, shot on 70mm, f / 8.0, --ar 2:3 --v 6.0

一位身穿蓝白相间运动服的中国高中女生坐在红色跑道上，背景是带有运动器材和看台的草地，她的头发扎在脑后，穿着运动鞋，精致的五官，16K，精确的 ISO 150 设置，焦距 70mm，光圈值 f / 8.0，出图比例 2∶3，版本 v 6.0

其他模板：

模板①提示词：A high school girl wearing high school uniform and a light gray skirt, in the classroom, sitting at her desk with an open notebook on it, next to two water bottles, outside of which can see green trees through large windows, with bright sunshine shining into the room, creating a warm atmosphere, professional color grading, soft shadows, 8K, precise ISO 150 settings, shot on 70mm, f / 8.0, --ar 2:3 --v 6.0

一位穿着高中校服和浅灰色的裙子的高中女生，在教室里，坐在放着一本打开的笔记本的书桌前，旁边是两个水瓶，从大窗户可以看到外面的绿树，明亮的阳光照进房间，营造出温暖的气氛，专业的色彩分级，柔和的阴影，8K，精确的ISO 150设置，焦距70mm，光圈值f / 8.0，出图比例2∶3，版本v 6.0

模板②提示词：Campus photo, campus playground, sunshine, a beautiful Chinese girl sitting on the ground reading, wearing a white school uniform and black tie, with two braids, in front of a football goal post, precise ISO 200 settings, shot on 50mm, f / 11, --ar 3:2 --v 6.0

校园照片，校园操场，阳光，一位美丽的中国女孩坐在地上读书，穿着白色的校服并打着黑色领带，扎着两条辫子，在足球门柱前，精确的ISO 200设置，焦距50mm，光圈值f / 11，出图比例3∶2，版本v 6.0

例：替换成模板①，如图9.7-2所示。

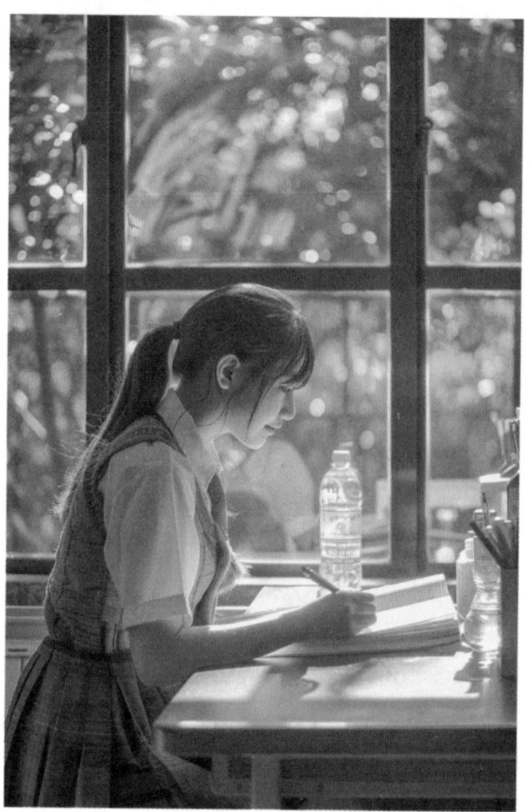

图9.7-2

Tips

在进行人像摄影时，相机的自动白平衡模式可能会受到环境因素的影响，如周围墙壁或脸部衣服上的反射光。在人像情况下，依赖自动白平衡可能会导致结果不一致，如果需要让主体发光，可以将可用光反射回主体脸上。

9.8 法式画报风格

法式画报风格效果图如图 9.8-1 所示。

图 9.8-1

主题：法式画报风格	软件：Midjourney	
光圈值：f / 8.0	感光度：ISO 150	焦距：85mm
色彩：暖色为主	光影：自然光	构图：中景

提示词：A young woman in early Victorian attire, wearing a light colored dress with lace trim and hat, hazy atmosphere, sitting on the grass of an ancient pond in central park, on a sunny day, photo realistic, soft natural lighting, intricate details of the and expression, old photograph style, 16K, precise ISO 150 settings, shot on 85mm, f / 8.0, --ar 2:3 --v 6.0

一位年轻女子身着维多利亚时代早期的服装，身穿浅色蕾丝边连衣裙、戴着帽子，气氛朦胧，坐在中央公园里老池塘旁的草地上，阳光明媚的日子，写实的照片，柔和的自然采光，错综复杂的细节和表情，老照片风格，16K，精确的 ISO 150 设置，焦距 85mm，光圈值 f / 8.0，出图比例 2∶3，版本 v 6.0

其他模板：

模板①提示词：Photography, maiden in the garden, French style, classical elegance, lush scenery, soft lights, asymmetrical composition, depth of field, 8K, precise ISO 150 settings, shot on 50mm, f / 8.0, --ar 2:3 --v 6.0

摄影，庭园中的少女，法式风格，古典优雅，风景葱郁，灯光柔和，构图不对称，景深，8K，精确的 ISO 150 设置，焦距 50mm，光圈值 f / 8.0，出图比例 2：3，版本 v 6.0

模板②提示词：Photo, a girl sitting on the windowsill, French style, long white dress, strappy sandals, long brown hair, smiling face, hanging clusters of flowers, lush scenery, asymmetrical composition, knee shot, classical elegance, depth of field, precise ISO 200 settings, shot on 50mm, f / 11, --ar 3:4 --v 6.0

照片，一个女孩坐在窗台上，法式风格，白色长裙，系带凉鞋，棕色长发，笑容满面，挂着一簇簇鲜花，景色郁郁葱葱，构图不对称，膝盖以上视角，古典优雅，景深，精确的 ISO 200 设置，焦距 50mm，光圈值 f / 11，出图比例 3：4，版本 v 6.0

例：替换成模板①，如图 9.8-2 所示。

图 9.8-2

Tips

在进行法式画报风格摄影时，更推荐采取柔和的光线进行拍摄。除此之外，还可以利用撞色，在柔和的光影中，让色彩形成一道鲜明的风景，但要记得选择低饱和度的色彩，才能打造出类似莫兰迪的画中的柔和色调。

9.9 日式小清新风格

日式小清新风格效果图如图 9.9-1 所示。

图 9.9-1

主题：日式小清新风格	软件：Midjourney	
光圈值：f / 8.0	感光度：ISO 150	焦距：85mm
色彩：暖色为主	光影：自然光	构图：中景

提示词：Japanese high school girl, with a black coat and skirt, cherry blossoms in the background, carrying a shoulder bag, full body photo, in the style of a Japanese artist, 16K, precise ISO 150 settings, shot on 85mm, f / 8.0, --ar 2:3 --v 6.0

日本高中女生，穿着黑色外套和裙子，背景是樱花，背着双肩包，全身照，日本艺术家的风格，16K，精确的 ISO 150 设置，焦距 85mm，光圈值 f / 8.0，出图比例 2∶3，版本 v 6.0

其他模板：

模板①提示词：A Chinese girl with black hair, bangs and shoulder length hair, wearing white, leaning on the fence, listening to music through headphones, she wears an off-white straw hat and has exquisite facial features, with a side view angle, 8K, precise ISO 150 settings, shot on 70mm, f / 8.0, --ar 2:3 --v 6.0

一位留着黑发、刘海和齐肩长发的中国女孩，穿着白色衣服，靠在栏杆上，戴着耳机听音乐，戴着一顶灰白色的草帽，有着精致的五官，侧视图，8K，精确的 ISO 150 设置，焦距 70mm，光圈值 f / 8.0，出图比例 2∶3，版本 v 6.0

模板②提示词：A girl with long black hair, wearing blue short sleeves and white skirt, on the bridge in summer, eyes closed and smiles, holds an old camera in one hand, the background features green trees and sky, creating a fresh atmosphere, the youthful vitality, precise ISO 150 settings, shot on 50mm, f / 11, --ar 3:4 --v 6.0

一位留着黑色长发的女孩，穿着蓝色短袖和白色裙子，在夏日的桥上，闭着眼睛微笑着，手里拿着一台旧相机，背景是绿色的树木和天空，营造出清新的氛围，青春的活力，精确的 ISO 150 设置，焦距 50mm，光圈值 f / 11，出图比例 3∶4，版本 v 6.0

例：替换成模板①，如图 9.9-2 所示。

图 9.9-2

Tips

拍摄日系的人像照片时，可以让色彩饱和度低一点，对比度也低一点。逆光的日系照片很漂亮，可以开大光圈，将太阳放在取景器外一点点，从而营造出逆光效果。

第10章　景别控制：不同角度看世界

景别是指拍摄画面在取景器内所呈现的范围大小，通常随镜头和被摄主体之间的距离改变。被摄主体离镜头越近，画面范围越小，景别越小；被摄主体离镜头越远，画面范围越大，景别就越大。如果想要收获不同的画面效果，就要学会利用不同情况下的景别设置。

10.1 微距镜头

微距镜头效果图如图 10.1–1 所示。

图 10.1–1

主题：微距镜头　　软件：Midjourney

光圈值：f / 8.0　　感光度：ISO 200　　焦距：100mm

色彩：暖色为主　　光影：灯光　　构图：特写

提示词：Macro photography of a ladybug on plants, dew drops and water droplets in the air, reflection, macro lens, professional studio lighting, professional color grading, high resolution photography, 8K, precise ISO 200 settings, shot on 100mm, f / 8.0, --ar 4:5 --v 6.0

植物上的瓢虫微距摄影，空气中的露珠和水滴，反射，微距镜头，专业工作室灯光，专业色彩分级，高分辨率摄影，8K，精确的 ISO 200 设置，焦距 100mm，光圈值 f / 8.0，出图比例 4：5，版本 v 6.0

其他模板：

模板①提示词：Macro photography, the drops of water on the petals, realist detail, high resolution photography, precise ISO 150 settings, shot on 90mm, f / 8.0, --ar 3:2 --v 6.0

微距摄影，花瓣上的水滴，写实的细节，高分辨率摄影，精确的 ISO 150 设置，焦距 90mm，光圈值 f / 8.0，出图比例 3：2，版本 v 6.0

模板②提示词：Macro photography, growing moss on rocks, green, fresh, blurred background, extreme closeup view, exaggerated proportions, visually striking, elegant lines, 4K, precise ISO 150 settings, shot on 16mm, f / 11, --ar 3:4 --v 6.0

微距摄影，岩石上生长的苔藓，绿色、清新、模糊的背景，极端特写视图，夸张的比例，具有视觉冲击力，优雅的线条，4K，精确的 ISO 150 设置，焦距 16mm，光圈值 f / 11，出图比例 3：4，版本 v 6.0

例：替换成模板①，如图 10.1–2 所示。

图 10.1–2

Tips

一般情况下，我们会用微距镜头拍摄昆虫、植物等，所以对微距摄影师来说，可以选择花草较多的地方，如植物园。在拍摄昆虫时，最好的外出室外温度应大于 17℃，因为昆虫的活跃程度随室外温度改变，当天气温暖时，昆虫会更活跃。除此之外，阴天通常比晴天效果好，因为阴天会发出柔和的光，不会让画面过曝。

10.2 鱼眼镜头

鱼眼镜头效果图如图 10.2-1 所示。

图 10.2-1

主题：鱼眼镜头　　软件：Midjourney

光圈值：f / 8.0　　感光度：ISO 200　　焦距：16mm

色彩：暖色和冷色相结合　　光影：灯光　　构图：全景

提示词：Fish-eye effect, the GoPro view of city view, photo, night view, tall buildings, overlooking view, city night views in the background, fine art cinematic portrait photography, impressive, street art sensibilities, realist detail, 8K, precise ISO 200 settings, shot on 16mm, f / 8.0, --ar 2:3 --v 6.0

鱼眼效果，GoPro 视图的城市景观，照片，夜景，高层建筑，俯瞰视图，城市夜景背景，美术电影肖像摄影，令人印象深刻，街头艺术感，现实主义细节，8K，精确的 ISO 200 设置，焦距 16mm，光圈值 f / 8.0，出图比例 2∶3，版本 v 6.0

其他模板：

模板①提示词：Fish-eye effect, photography, the GoPro view of a girl walking down the street, in the bustling street, city night views in the background, fine art cinematic portrait photography, street art, realist detail, precise ISO 150 settings, shot on 16mm, f / 11, --ar 2:3 --v 6.0

鱼眼效果，摄影，GoPro 拍摄走在街上的女孩，在繁华的街道上，背景是城市夜景，精美的艺术电影肖像摄影，街头艺术感，现实主义的细节，精确的 ISO 150 设置，焦距 16mm，光圈值 f / 11，出图比例 2∶3，版本 v 6.0

模板②提示词：Fish-eye effect, photography, the GoPro view of a cute and beautiful girl, girl pouncing to viewer, extreme close up perspective, focus on face, extreme wide angle, city background, high definition, rich texture, highlight performance, 4K, precise ISO 150 settings, shot on 16mm, f / 11, --ar 3:4 --v 6.0

鱼眼效果，摄影，GoPro 拍摄的一个可爱美丽的女孩，女孩扑向观众，极端近距离视角，专注于脸部，极端广角，城市背景，高清晰度，丰富的纹理，突出表现，4K，精确的 ISO 150 设置，焦距 16mm，光圈值 f / 11，出图比例 3∶4，版本 v 6.0

例：替换成模板①，如图 10.2-2 所示。

图 10.2-2

Tips

使用鱼眼镜头时需要注意，越接近画面边缘的线条，变形程度就越大。如果把地平线安排在画面中央，那么最终呈现的效果会是水平的。只有把地平线安排在画面边缘，才能呈现出夸张的变形程度。与之相对的，人物应放在画面中心。

10.3 广角镜头

广角镜头效果图如图 10.3-1 所示。

图 10.3-1

主题：广角镜头	软件：Midjourney	
光圈值：f / 8.0	感光度：ISO 200	焦距：30mm
色彩：暖色为主	光影：自然光	构图：全景

提示词：Wide-angle lens, ancient Chinese architecture, the red walls and gray tiles shining under sunset light, stands on an ice-covered lake surface, surrounded by trees swaying gently in the breeze, the sky is painted with vibrant colors of purple, blue, pink, orange, and yellow, serene atmosphere, high resolution, 8K, precise ISO 200 settings, shot on 30mm, f / 8.0, --ar 2:1 --v 6.0

广角镜头，中国古代建筑，红色的墙壁和灰色的瓷砖在夕阳下闪闪发光，矗立在冰封的湖面上，周围的树木在微风中轻轻地摇曳，天空被涂上了紫色、蓝色、粉红色、橙色和黄色等鲜艳的颜色，宁静的氛围，高分辨率，8K，精确的 ISO 200 设置，焦距 30mm，光圈值 f / 8.0，出图比例 2：1，版本 v 6.0

其他模板：

模板①提示词：The vast and endless salt lake of the Dead Sea, with clear water, a few people standing by it at sunset, reflections of clouds on top, high definition, highlighting the beauty of nature's scenery, bright colors, highlight the contrast between light and dark, precise ISO 150 settings, shot on 24mm, f / 11, --ar 2:1 --v 6.0

死海广阔无边的盐湖，湖水清澈，日落时几个人站在湖边，云的倒影在上面，高清摄影，凸显自然风光之美，亮色，突出明暗对比，精确的 ISO 150 设置，焦距 24mm，光圈值 f / 11，出图比例 2：1，版本 v 6.0

模板②提示词：A team of mountaineers climbing the Himalayas, surrounded by snow-covered peaks and icy cliffs, mountain textures and distant clouds, soft natural lighting, with rocks in the foreground, some hikers far away in the background, 4K, precise ISO 150 settings, shot on 35mm, f / 11, --ar 3:4 --v 6.0

登山队攀登喜马拉雅山脉，周围是白雪覆盖的山峰和冰冷的悬崖，山的纹理和远处的云，柔和的自然光，前景中有岩石，远处的背景中有一些徒步旅行者，4K，精确的 ISO 150 设置，焦距 35mm，光圈值 f / 11，出图比例 3：4，版本 v 6.0

例：替换成模板①，如图 10.3-2 所示。

图 10.3-2

Tips

因为广角镜头会扭曲视角，让画面的主体元素显得很远，所以可以设置几个前置元素来增加纵深感，有时候，前景元素可以创建出引导线，将观者的视线引至画面的主体位置。

10.4 无人机俯视镜头

无人机俯视镜头效果图如图 10.4-1 所示。

图 10.4-1

主题：无人机俯视镜头　　软件：Midjourney

光圈值：f / 8.0　　感光度：ISO 150　　焦距：24mm

色彩：暖色和冷色相结合　　光影：环境光　　构图：远景

提示词：Photo, the aerial view, drone shooting, the mountain road, the bright colors, there were red trees on the left and green trees on the right, and a road ran between them, 8K, precise ISO 150 settings, shot on 24mm, f / 8.0, --ar 3:4 --v 6.0

照片，鸟瞰图，无人机拍摄，山路，明亮的色彩，左边是红色的树、右边是绿色的树，中间穿插着一条公路，8K，精确的 ISO 150 设置，焦距 24mm，光圈值 f / 8.0，出图比例 3 : 4，版本 v 6.0

其他模板：

模板①提示词：Photography, aerial view, drone shooting, canyon, river running through the middle of the canyon on both sides, natural light, precise ISO 150 settings, shot on 24mm, f / 11, --ar 3:4 --v 6.0

摄影，鸟瞰图，无人机拍摄，峡谷，峡谷两边的河流，自然光，精确的 ISO 150 设置，焦距 24mm，光圈值 f / 11，出图比例 3 : 4，版本 v 6.0

模板②提示词：Landscape photography, an aerial view of the vast expanse of lake, with an endless sea and green grassland stretching to the horizon, a white car is driving on the road leading towards the lake, serene scene, the blue sky above contrasts beautifully against the deep water below, 4K, precise ISO 150 settings, shot on 35mm, f / 8.0, --ar 3:4 --v 6.0

风景摄影，鸟瞰视角下辽阔的湖面，一望无际的大海和绿色草原，一辆白色的汽车行驶在通往湖边的路上，宁静的景色，上面的蓝天与下面的海水形成了美丽的对比，4K，精确的 ISO 150 设置，焦距 35mm，光圈值 f / 8.0，出图比例 3 : 4，版本 v 6.0

例：替换成模板①，如图 10.4-2 所示。

图 10.4-2

Tips

利用无人机航拍时，在飞行前，需要对航拍区域进行全面的了解。其中包括地形、天气、植被、人员和建筑，以及限飞高度等情况，从而做好安全飞行和拍摄计划。需要选择远离障碍物和高压线等危险区域的安全航线。同时，在无人机起飞和降落时，需要选择宽敞、平整、远离人群的区域。

10.5 特写镜头

特写镜头效果图如图 10.5-1 所示。

图 10.5-1

主题：特写镜头	软件：Midjourney	
光圈值：f / 8.0	感光度：ISO 200	焦距：70mm
色彩：冷色为主	光影：自然光	构图：特写

提示词：A beautiful girl with big eyes, covered in snow and wearing thick winter, is looking at the camera with an expression of curiosity, focusing on her face, close-up view, delicate facial features, dark hair that shines under the sunlight, 8K, precise ISO 200 settings, shot on 70mm, f / 8.0, --ar 2:3 --v 6.0

一位漂亮的大眼睛女孩，裹着雪且穿着厚厚的冬衣，带着好奇的表情看着镜头，焦点集中在她的脸上，特写镜头，精致的五官，黑色的头发在阳光下闪闪发光，8K，精确的 ISO 200 设置，焦距 70mm，光圈值 f / 8.0，出图比例 2 ∶ 3，版本 v 6.0

其他模板：

模板①提示词：A close up view of a tiger's eye with orange colored eyes, single eye, macro zoom, local close-up, realistic light and shadow, hyper-realistic animal illustrations, vivid, full vibrant, highly detailed, precise ISO 150 settings, shot on 100mm, f / 11, --ar 5:3 --v 5.2

老虎橙色眼睛的特写，单眼，微距变焦，局部特写，逼真的光影，超现实的动物插图，生动，充满活力，高度细致，精确的 ISO 150 设置，焦距 100mm，光圈值 f / 11，出图比例 5 ∶ 3，版本 v 5.2

模板②提示词：Close up of a woman with long hair blowing in the wind, wrapped around her neck is wearing white ribbons, sun rays coming through an open window, dreamy, ethereal, cinematic, precise ISO 200 settings, shot on 80mm, f / 11, --ar 2:3 --v 6.0

一位被风吹拂长发的女人的特写，脖子上缠着白色的丝带，阳光从敞开的窗户照进来，梦幻的、缥缈的、电影般的，精确的 ISO 200 设置，焦距 80mm，光圈值 f / 11，出图比例 2 ∶ 3，版本 v 6.0

例：替换成模板①，如图 10.5-2 所示。

图 10.5-2

Tips

想让局部特写好看，色彩运用是非常重要的一环，不单是画面上的表现，还包含了不同的情绪和意义。例如，以红、橙、黄为代表的暖色调，能为画面带来温暖的感受，营造出热情、积极的情绪；以蓝、绿为代表的冷色调，则有一种平静、安宁的感觉；而两种色调同时出现在画面里，又能产生对立、矛盾、戏剧性的视觉效果，从而让摄影作品更加醒目有趣。

10.6 远景拍摄

远景拍摄效果图如图 10.6–1 所示。

图 10.6–1

主题：远景拍摄	软件：Midjourney	
光圈值：f / 8.0	感光度：ISO 200	焦距：35mm
色彩：冷色为主	光影：自然光	构图：远景

提示词：In the glacier lagoon, an icy wonderland of cool blues and whites, a person stands on one block looking up at endless fields of floating snow–covered ice cubes, a surreal scene, a wide–angle lens to capture every detail of these mesmerizing frozen landscapes, 8K, precise ISO 200 settings, shot on 35mm, f / 8.0, ––ar 2:3 ––v 6.0

冰川潟湖，一个冷蓝色和白色的冰雪仙境，一个人站在一个石块上、仰望着一望无际的被白雪覆盖的漂浮冰块，一种超现实的场景，使用广角镜头捕捉到这些迷人的冰景的细节，8K，精确的 ISO 200 设置，焦距 35mm，光圈值 f / 8.0，出图比例 2∶3，版本 v 6.0

其他模板：

模板①提示词：Long shot, the background is the snow mountain, with green grasslands and winding roads in front, a girl runs on an empty road, in the style of nature, realist detail, precise ISO 150 settings, shot on 35mm, f / 11, ––ar 4:5 ––v 6.0

远景，背景是雪山，前面是绿色的草原和蜿蜒的道路，一个女孩在空旷的道路上奔跑，自然的风格，现实主义的细节，精确的 ISO 150 设置，焦距 35mm，光圈值 f / 11，出图比例 4∶5，版本 v 6.0

模板②提示词：Long shot, a wooden bridge is walking along the autumn maple leaves covered road, realistic blue skies, strong light effect, ultra wide shot, extreme angle, 32K, precise ISO 100 settings, shot on 24mm, f / 11, ––ar 16:9 ––v 6.0

远景，一座木桥沿着秋天的枫叶覆盖的道路，逼真的蓝天，强烈的光效，超广角镜头，极端的角度，32K，精确的 ISO 100 设置，焦距 24mm，光圈值 f / 11，出图比例 16 : 9，版本 v 6.0

例：替换成模板①，如图 10.6-2 所示。

图 10.6-2

Tips

通常情况下，在进行远景拍摄时，画面构图不会用到前景，而是着重通过深远的景物和开阔的视野将观众的视线引向远方。所以，拍摄过程中尽量不用顺光，而是选择侧光或者侧逆光以形成画面层次，突出透视效果。在此基础上，还要注意画面远处的透视关系及影调的明暗，避免画面单调乏味。

10.7 空镜拍摄

空镜拍摄效果图如图 10.7-1 所示。

图 10.7-1

主题：空镜拍摄　　软件：Midjourney
光圈值：f / 8.0　　感光度：ISO 200　　焦距：50mm
色彩：暖色为主　　光影：自然光　　构图：中景

提示词：A beautiful photograph of white daisies growing in a field, with a clear blue sky in the background, scenery shot, 8K, precise ISO 200 settings, shot on 50mm, f / 8.0, --ar 3:2 --v 6.0

一张白色雏菊在田野里生长的美丽照片，背景是湛蓝的天空，空镜头，8K，精确的ISO 200设置，焦距50mm，光圈值f / 8.0，出图比例3∶2，版本v 6.0

其他模板：

模板①提示词：The foreground features a closeup of fern leaves, with some light and shadow effects, minimalist style, with soft tones and natural elements, with soft colors and tones, scenery shot, 4K, precise ISO 150 settings, shot on 35mm, f / 11, --ar 3:2 --v 6.0

前景以蕨类植物的叶子特写为特色，有一些光影效果，极简主义，柔和的色调和自然元素，柔和的色彩与色调，空镜，4K，精确的ISO 150设置，焦距35mm，光圈值f / 11，出图比例3∶2，版本v 6.0

模板②提示词：Roses placed next to tombstones, creating an atmosphere full of sadness, the background has gray cement slabs and flowers scattered around them, a subtle texture that resembles dirt or sand, outdoors at night time, melancholic feeling, scenery shot, 16K, precise ISO 150 settings, shot on 35mm, f / 11, --ar 3:2 --v 6.0

墓碑旁摆放着玫瑰花，营造出充满悲伤的气氛，背景有灰色水泥板和周围散落的花朵，类似泥土或沙子的微妙质感，在夜晚的户外，忧郁的感觉，空镜，16K，精确的ISO 150设置，焦距35mm，光圈值f / 11，出图比例3∶2，版本v 6.0

例：替换成模板①，如图10.7-2所示。

图10.7-2

Tips

空镜是指视频中仅展示自然景物或细节，而不出现和剧情有关的人物的镜头，可以放在影片开头，介绍整个故事的背景环境，也常用来侧面表达人物的情感。例如，影片中常用升起的太阳象征希望，用雨天传递人物失落的心情等。所以空镜的拍摄需要结合剧情的发展而定。

10.8 近景拍摄

近景拍摄效果图如图 10.8-1 所示。

图 10.8-1

主题：近景拍摄　　**软件：**Midjourney

光圈值：f / 8.0　　**感光度：**ISO 200　　**焦距：**100mm

色彩：暖色为主　　**光影：**自然光　　**构图：**近景

提示词：Close shot, in the background are the blurred autumn leaves, beautiful girl in a white dress with brown hair in the sun with soft light, portrait closeup in the style of a dreamy style, photo with high resolution, 8K, precise ISO 200 settings, shot on 100mm, f / 8.0, --ar 2:3 --v 6.0

近景，背景是模糊的秋叶，在柔和的阳光下穿着白色连衣裙的美丽棕发女孩，梦幻风格的肖像特写，高分辨率的照片，8K，精确的 ISO 200 设置，焦距 100mm，光圈值 f / 8.0，出图比例 2∶3，版本 v 6.0

其他模板：

模板①提示词：A photo of an Asian beauty girl standing under cherry trees, wearing white with pink flowers on it, in the sunshine, with a light blue background, high resolution, close shot, 4K, precise ISO 150 settings, shot on 85mm, f / 11, --ar 2:3 --v 6.0

一张亚洲美女站在樱花树下的照片，穿着白色的衣服、上面有粉红色的花，在阳光下，浅蓝色的背景，高分辨率，近景，4K，精确的 ISO 150 设置，焦距 85mm，光圈值 f / 11，出图比例 2∶3，版本 v 6.0

模板②提示词：A beautiful Chinese girl with long hair, wearing an off-the-shoulder blue sweater and white skirt, in the garden under cherry blossoms, smiling, wearing a straw hat on her head, with very delicate makeup, exquisite facial features, close up, portrait photography, 16K, precise ISO 150 settings, shot on 85mm, f / 11, --ar 3:4 --v 6.0

一位留着长发的美丽中国女孩，穿着露肩蓝毛衣和白裙子，在樱花盛开的花园，微笑着，头上戴着草帽，妆容精致，五官精致，特写，人像摄影，16K，精确的ISO 150设置，焦距85mm，光圈值f/11，出图比例3:4，版本v 6.0

例：替换成模板①，如图10.8-2所示。

图10.8-2

Tips

近景拍摄的特点是景别小、内容少、亮度变化不复杂。在拍摄时，要注意测光和曝光，当拍摄的主体是人物时，一般按人脸的亮度值确定曝光，使人物在画面中有较好的表现。近景画面着重强调色彩还原和神情表达，因此光线效果需要突出人脸或某一局部。除此之外，还要考虑到背景与人物之间的色彩协调与搭配。

10.9 全景拍摄

全景拍摄效果图如图10.9-1所示。

图10.9-1

主题：全景拍摄　　软件：Midjourney
光圈值：f / 8.0　　感光度：ISO 200　　焦距：35mm
色彩：暖色为主　　光影：自然光　　构图：全景

提示词：Panoramic photography, a silhouette of an athlete standing on top of the hill at sunset, holding their gear in hand, looking out over the horizon, the sky is painted in shades of pink and orange, evening light bathes the scene, 16K, precise ISO 200 settings, shot on 35mm, f / 8.0, --ar 3:2 --v 6.0

全景摄影，一名运动员站在日落时分山顶上的剪影，手里拿着装备，望着地平线，天空被涂上了粉红色和橙色的阴影，黄昏的光线沐浴在场景中，16K，精确的 ISO 200 设置，焦距 35mm，光圈值 f / 8.0，出图比例 3：2，版本 v 6.0

其他模板：

模板①提示词：Panoramic photography, full body, a beautiful woman in white dress, stands on the field, blue sky with clouds, in front her is a vast expanse of grassland, with several straw, a gentle breeze blows through her hair, 4K, precise ISO 150 settings, shot on 35mm, f / 11, --ar 3:2 --v 6.0

全景摄影，全身，一位穿着白色衣服的美丽女人，站在平原上，蓝天白云，前面是一片广阔的草原，有些稻草，微风吹过她的头发，4K，精确的 ISO 150 设置，焦距 35mm，光圈值 f / 11，出图比例 3：2，版本 v 6.0

模板②提示词：The aerial view of the ancient city, panoramic photography, an old red tile building with yellow walls and green tiles, surrounding environment has trees and grassland, some white metal guardrail on both sides, 16K, precise ISO 150 settings, shot on 35mm, f / 11, --ar 3:2 --v 6.0

古城鸟瞰图，全景摄影，一座外墙是黄色的、屋顶是绿色瓦片的旧红色砖瓦建筑，周围环境有树木和草地，两边有一些白色的金属护栏，16K，精确的 ISO 150 设置，焦距 35mm，光圈值 f / 11，出图比例 3 ：2，版本 v 6.0

例：替换成模板①，如图 10.9-2 所示。

图 10.9-2

Tips

全景通常采用加宽的画面长度来记录场景，大范围的横向取景宽度可以充分展现出场景的宏伟、辽阔，从而达到观看者对场景整体景观一目了然的目的。

10.10 中景拍摄

中景拍摄效果图如图 10.10–1 所示。

图 10.10–1

主题：中景拍摄　　**软件：Midjourney**

光圈值：f / 8.0　　**感光度：ISO 200**　　**焦距：50mm**

色彩：冷色为主　　**光影：自然光**　　**构图：中景**

提示词：Medium shot, a photograph of an Asian woman leaning against the wall aboard ship, with black hair and glasses, wearing a black shirt, the sea is rough with white foam, pale colors with dark contrast, shot from above in a close up, 8K, precise ISO 200 settings, shot on 50mm, f / 8.0, --ar 2:3 --v 6.0

中景，一张亚洲女人在船上靠着墙的照片，黑发并戴着眼镜，穿着黑色衬衫，海面波涛汹涌且有白色的泡沫，浅色与深色对比，从上方近距离拍摄，8K，精确的 ISO 200 设置，焦距 50mm，光圈值 f / 8.0，出图比例 2 : 3，版本 v 6.0

其他模板：

模板①提示词：Medium shot, a beautiful woman holding flowers, wearing a white sweater, standing against a backdrop of green trees, with a yellow autumn forest in the background, focusing on capturing her expression while she holds the bouquet, 4K, precise ISO 150 settings, shot on 50mm, f / 11, --ar 2:3 --v 6.0

中景，一位手持鲜花的美丽女子，身穿白色毛衣，站在绿树的背景下，背景是一片黄色的秋林，重点捕捉她手持花束时的表情，4K，精确的 ISO 150 设置，焦距 50mm，光圈值 f/11，出图比例 2：3，版本 v 6.0

模板②提示词：Photo, medium shot, a beautiful Chinese girl with long hair is sitting on the lawn, in a white short-sleeved outfit, she has two braided ponytails, 16K, precise ISO 150 settings, shot on 50mm, f/11, --ar 3:4 --v 6.0

照片，中景，一位美丽的长发中国女孩坐在草坪上，穿着白色短袖衣服，有两个扎成辫子的马尾，16K，精确的 ISO 150 设置，焦距 50mm，光圈值 f/11，出图比例 3：4，版本 v 6.0

例：替换成模板①，如图 10.10-2 所示。

图 10.10-2

Tips

中景较全景而言，画面中的人物形象和背景环境并没有那么重要，而是更重视人物的形体动作、情绪交流和画面的情节。中景构图有利于交代人物和背景之间的关系，并且可以反映出较强的画面结构和人物交流区域。

第11章　特效摄影：超有趣的画面体验

在摄影的过程中，如果想要达到一些非现实的艺术效果，可以在前期利用特效镜头、照明、布景，人工制造出一种假象和幻觉，这就是特效摄影。在电影制作行业中，特效摄影可以避免让演员处于危险的境地，同时减少制作成本，增加艺术效果。

11.1 红外效果

红外效果图如图 11.1-1 所示。

图 11.1-1

主题：红外效果	软件：Midjourney	
光圈值：f / 11	感光度：ISO 200	焦距：50mm
色彩：冷色为主	光影：环境光	构图：全景

提示词：An infrared photographic image, an elegant bridge over the river, weeping willow trees on both sides, the water with swans swimming in it, soft focus, dreamy and surreal style, high detailed style, depth of field, a fantasy atmosphere, 32K, precise ISO 200 settings, shot on 50mm, f / 11, --ar 3:2 ---v 6.0

一张红外摄影图像，河上有一座优雅的桥，两边有垂柳，天鹅在河里游泳，软焦点，梦幻和超现实的风格，高度细致的风格，景深，一种梦幻般的氛围，32K，精确的 ISO 200 设置，焦距 50mm，光圈值 f / 11，出图比例 3：2，版本 v 6.0

其他模板：

模板①提示词：An infrared photographic image, a big tree on a grassy hillside, natural lighting, asymmetrical composition, spectacular backdrops, realistic yet romantic scenery, photo-realistic techniques, 8K, precise ISO 150 settings, shot on 35mm, f / 11,--ar 16:9 --v 5.2

一张红外摄影图像，长满草的山坡上的一棵树，自然采光，不对称构图，壮观的背景，现实而浪漫的风景，逼真的技术，8K，精确的 ISO 150 设置，焦距 35mm，光圈值 f / 11，

出图比例 16∶9，版本 v 5.2

模板②提示词：Infrared photography, the serene beauty of an enchanted winter landscape, delicate frost covered trees and wild grasses, a peaceful scene, precise ISO 100 settings, shot on 35mm, f / 8.0, --ar 3:4 --v 6.0

红外摄影，宁静、美丽、迷人的冬季景观，精致的霜冻覆盖着树木和野草，一个宁静的场景，精确的 ISO 100 设置，焦距 35mm，光圈值 f / 8.0，出图比例 3∶4，版本 v 6.0

例：替换成模板①，如图 11.1–2 所示。

图 11.1–2

Tips

红外摄影是指将原本在自然景观中可见的普通原色，通过红外技术进行转化，形成另一种神奇的透视效果，为观众打造出全新的视觉感受。

11.2 光绘效果

光绘效果图如图 11.2–1 所示。

图 11.2–1

主题：光绘效果	**软件：**Midjourney	
光圈值：f / 8.0	**感光度：**ISO 100	**焦距：**24mm
色彩：冷色为主	**光影：**自然光	**构图：**远景

提示词：Light painting photography, night sky with crescent moon and stars above rocky beach in far distance, sparks flying from small fire on rock near water's edge, distant silhouette of tall cliff building structure at horizon, 16K, precise ISO 100 settings, shot on 24mm, f / 8.0, --ar 3:2 --v 6.0

光绘摄影，远处有岩石、沙滩、新月和星星的夜空，水边岩石上的小火喷出的火花，远处地平线上高大的悬崖建筑结构的剪影，16K，精确的 ISO 100 设置，焦距 24mm，光圈值 f / 8.0，出图比例 3∶2，版本 v 6.0

其他模板：

模板①提示词：Light painting photography, a beautiful bride in a white dress standing inside an illuminated circle, surrounded by sparks and light streaks in the style of light painting, in the desert at night, dark blue sky, 4K, precise ISO 150 settings, shot on 24mm, f / 8.0, --ar 3:2 --v 6.0

光绘摄影，一位美丽的新娘身穿白色礼服站在一个被照亮的圆圈里，周围是光绘风格的火花和光条，在夜晚的沙漠里，深蓝色的天空，4K，精确的 ISO 150 设置，焦距 24mm，光圈值 f / 8.0，出图比例 3∶2，版本 v 6.0

模板②提示词：Light drawing renderings of the circle, light painting, dark orange and light gold, spirals, rule of thirds composition, colorful explosions, splash the sparks, night photography, hyper quality, precise ISO 200 settings, shot on 35mm, f / 8.0, --ar 5:3 --v 6.0

圆形光绘效果图，光绘，深橙色和浅金色，螺旋，三分法构图，色彩缤纷的爆炸，飞溅的火花，夜间摄影，超高品质，精确的 ISO 200 设置，焦距 35mm，光圈值 f / 8.0，出图比例 5∶3，版本 v 6.0

例：替换成模板①，如图 11.2-2 所示。

图 11.2-2

Tips

通常情况下，大部分光绘摄影作品是在全黑的室内或夜间弱光环境下完成的。但如果想要进行创新，除了光绘主体的光源之外，同样可以利用环境光（如月光、灯光等）为画面增加氛围感。

11.3 长焦效果

长焦效果图如图 11.3–1 所示。

图 11.3–1

主题：长焦效果	软件：Midjourney	
光圈值：f / 8.0	感光度：ISO 100	焦距：150mm
色彩：暖色为主	光影：自然光	构图：近景

提示词：Hummingbird is flying towards a pink flower, long focus effect, as if in front of eyes, blurred background, meticulous technique, nature's wonder, captivating, 16K, precise ISO 100 settings, shot on 150mm, f / 8.0, --ar 5:3 --v 5.2

蜂鸟正飞向一朵粉红色的花，长焦效果，仿佛就在眼前，模糊的背景，细致的技术，大自然的奇迹，迷人的，16K，精确的 ISO 100 设置，焦距 150mm，光圈值 f / 8.0，出图比例 5：3，版本 v 5.2

其他模板：

模板①提示词：Real photo, the moon is in the sky, and there's an electric tower on one side of it, at night with stars shining brightly above the blue sky, high definition, long focus effect, 16K, precise ISO 250 settings, shot on 135mm, f / 5.6, --ar 2:3 --v 6.0

真实照片，在天上的月亮，月亮的一侧有一座电塔，星星在蓝天上闪闪发光的夜晚，高清晰度，长焦效果，16K，精确的 ISO 250 设置，焦距 135mm，光圈值 f / 5.6，出图比例 2：3，版本 v 6.0

模板②提示词：A blue bird, flying and carrying berries in its beak, the backdrop of nature's beauty, show the vibrant colors of both birds and foliage, highlights their graceful movements as they playfully twirl through air, precise ISO 250 settings, shot on 150mm, f / 8.0, --ar 3:4 --v 6.0

一只蓝色的鸟，飞翔的同时嘴里叼着浆果，背景是自然的美景，展示了鸟类和树叶的鲜艳色彩，创造了一个视觉上令人惊叹的场景，突出了它们在空中嬉戏旋转的优雅动作，精确的 ISO 250 设置，焦距 150mm，光圈值 f / 8.0，出图比例 3：4，版本 v 6.0

例：替换成模板①，如图 11.3-2 所示。

图 11.3-2

Tips

长焦镜头的视角很小，拍摄时对相机的抖动非常敏感。因此，为了保证画面的清晰度，通常情况下不推荐手持相机拍摄，将相机固定在三脚架上，才能确保拍出高清的画面。

11.4 双重曝光效果

双重曝光效果图如图 11.4-1 所示。

图 11.4-1

主题：双重曝光效果	软件：Midjourney	
光圈值：f / 5.6	感光度：ISO 200	焦距：50mm
色彩：暖色为主	光影：自然光	构图：中景

提示词：A double exposure of the silhouette of an Asian woman, in white dress and flowing hair, against the backdrop of sunset sky, golden hour light, the background is blurred with soft edges, creating depth, warm tones and soft lighting enhance the romantic atmosphere, 8K, precise ISO 200 settings, shot on 50mm, f / 5.6, --ar 2:3 --v 5.2

双重曝光的亚洲女人的剪影，白色的裙子和飘逸的头发，背景是夕阳的天空，黄金时刻的光线，背景是模糊的软边缘，创造景深，温暖的色调与柔和的灯光增强了浪漫的气氛，8K，精确的 ISO 200 设置，焦距 50mm，光圈值 f / 5.6，出图比例 2∶3，版本 v 5.2

其他模板：

模板①提示词：Double exposure of a beautiful girl and flowers, captivating gaze, profile, light gray and amber, dreamy collages, layered landscapes, captures the essence of nature, 16K, precise ISO 150 settings, shot on 70mm, f / 5.6, --ar 2:3 --v 6.0

一位美丽的女孩和花朵的双重曝光图，迷人的凝视，侧面，浅灰色和琥珀色，梦幻的拼贴画，分层的景观，捕捉自然的本质，16K，精确的 ISO 150 设置，焦距 70mm，光圈值 f / 5.6，出图比例 2∶3，版本 v 6.0

模板②提示词：Double exposure of an attractive young man looking to his right side, background is cityscape and skyscrapers in black and white, he stands against the backdrop of modern buildings, flying birds in the sky, retro futuristic sci-fi photo style, precise ISO 250 settings, shot on 50mm, f / 8.0, --ar 3:4 --v 6.0

一位迷人的年轻人看着他的右侧的双重曝光效果照，背景是黑白色的城市景观和摩天大楼，他站在现代建筑的背景下，天上有飞鸟，复古未来科幻照片风格，精确的 ISO 250 设置，焦距 50mm，光圈值 f / 8.0，出图比例 3∶4，版本 v 6.0

例：替换成模板①，如图 11.4-2 所示。

图 11.4-2

Tips

拍摄双重曝光效果照片时，为了提高画面的意境效果，需要让人物稍微带点情绪。一般情况下，会选择拍摄半身或特写肖像，并选择拍摄侧脸或 3/4 侧脸，引导拍摄对象做出一些类似沉思的表情。而背景图一般选用风景类的图片，如森林、花丛、海上日出等。

11.5 色彩焦点

色彩焦点效果图如图 11.5-1 所示。

图 11.5-1

主题：色彩焦点	软件：Midjourney	
光圈值：f / 5.6	感光度：ISO 200	焦距：35mm
色彩：暖色为主	光影：自然光	构图：全景

提示词：A photo at sunset, it was dark except for the man in the middle who was wearing red, people were weaving through the crowded and noisy streets in a shadow, diagonal composition, low exposure, strong contrast of light and shadow, focusing on the people, juxtaposition of light and shadow, chiaroscuro, color focus style, raw street photography, matte photo, 8K, precise ISO 200 settings, shot on 35mm, f / 5.6, --ar 2:3 --v 6.0

一张日落时的照片，除了中间穿红色衣服的人外，周围一片黑暗，人们在阴影中穿行于拥挤嘈杂的街道，对角构图，低曝光，强烈的光影对比，对人聚焦，光影并置，明暗对比，彩色聚焦风格，原始街头摄影，哑光照片，8K，精确的 ISO 200 设置，焦距 35mm，光圈值 f / 5.6，出图比例 2∶3，版本 v 6.0

其他模板：

模板①提示词：Red roses lay down on cement in a shadow, complete, beautiful leaf, diagonal composition, low exposure, strong contrast of light and shadow, focusing on the leaf, color focus style, matte photo, 16K, precise ISO 150 settings, shot on 50mm, f / 5.6, --ar 16:9 --v 6.0

红玫瑰平放在水泥地上形成的阴影，完整的、美丽的叶子，对角构图，低曝光，强烈的光影对比，聚焦在叶子上，彩色聚焦风格，哑光照片，16K，精确的 ISO 150 设置，焦距 50mm，光圈值 f / 5.6，出图比例 16∶9，版本 v 6.0

模板②提示词：A person holding an umbrella is walking under a street lamp, surrounded by traffic, only the person holding an umbrella is wearing bright clothes, it was dark all around, focusing on the people color focus style, matte photo, 16K, precise ISO 100 settings, shot on 50mm, f / 5.6, --ar 4:3 --v 6.0

一个撑着伞走在路灯下的人，周围是川流不息的车辆，只有撑着伞的人穿着亮色的衣服，周围都是黑暗的，聚焦在人上，彩色聚焦风格，哑光照片，16K，精确的 ISO 100 设置，焦距 50mm，光圈值 f / 5.6，出图比例 4∶3，版本 v 6.0

例：替换成模板①，如图 11.5-2 所示。

图 11.5-2

Tips

拍摄色彩焦点图的关键是画面中两种反差大的颜色之间的对比，通过合理的构图安排，可以进一步突出距离，强化空间。还需要注意的一点是，互相对比的色彩，只有在饱和度最高时对比效果才最明显。如果色彩变浅或变深，两种颜色之间的对比效果便会减弱，达不到我们想要的视觉效果。

11.6 散景效果

散景效果图如图 11.6–1 所示。

图 11.6–1

主题：散景效果	软件：Midjourney	
光圈值：f / 8.0	感光度：ISO 200	焦距：70mm
色彩：暖色为主	光影：灯光	构图：中景

提示词：Bokeh effect, the wind chimes under the eaves, dark amber and gold, summer, faint spot of light, close–up shots, 8K, precise ISO 200 settings, shot on 70mm, f / 8.0, ––ar 2:3 ––v 5.2

散景效果，屋檐下的风铃，暗琥珀和金色，夏天，微弱的光斑，特写镜头，8K，精确的 ISO 200 设置，焦距 70mm，光圈值 f / 8.0，出图比例 2：3，版本 v 5.2

其他模板：

模板①提示词：Bokeh effect, a street Christmas tree, dark amber and gold, winter night, faint spot of light, close–up shots, festive atmosphere, 8K, precise ISO 150 settings, shot on 70mm, f / 8.0, ––ar 16:9 ––v 5.2

散景效果，街头圣诞树，暗琥珀和金色，冬夜，微弱的亮点，特写镜头，节日气氛，8K，精确的 ISO 150 设置，焦距 70mm，光圈值 f / 8.0，出图比例 16：9，版本 v 5.2

模板②提示词：A dandelion in the foreground, bokeh effect, dark grey background, light white and silver style, soft edges and blurred details, blurred surrealism, minimalist naturalistic photography, precise ISO 250 settings, shot on 50mm, f / 8.0, ––ar 3:4 ––v 6.0

前景中的蒲公英，散景效果，深灰色的背景，浅白色和银色的风格，柔和的边缘和模糊的细节，模糊的超现实主义，极简的自然主义摄影，精确的 ISO 250 设置，焦距 50mm，光圈值 f / 8.0，出图比例 3 ：4，版本 v 6.0

例：替换成模板①，如图 11.6-2 所示。

图 11.6-2

Tips

要拍出散景光斑，长焦定焦镜头是首选，建议焦距至少有 50mm 或以上，这样的长焦距才会把主体与背景的距离压缩，模糊背景、突出光斑效果主体。如果只有短焦距镜头，那么一定要控制并保证主体和背景间有一定的距离，主体到相机的距离必须远远小于主体到背景的距离。

11.7 动态模糊效果

动态模糊效果图如图 11.7-1 所示。

图 11.7-1

主题：动态模糊效果	**软件**：Midjourney	
光圈值：f / 8.0	**感光度**：ISO 200	**焦距**：35mm
色彩：暖色为主	**光影**：灯光	**构图**：全景

提示词：A man rides a motorcycle in the city, motion blur photography, night scene, movie lighting, low angle shot, high speed shutter, neon lights illuminate streets, bustling traffic and pedestrians passing by, long exposure, high resolution, best quality, super detail, professional color grading, 8K, precise ISO 200 settings, shot on 35mm, f / 8.0, --ar 3:2 --v 6.0

一个在城市里骑着摩托车的人，动态模糊摄影，夜景，电影灯光，低角度拍摄，高速快门，霓虹灯照亮街道，熙熙攘攘的交通和行人经过，曝光时间长，高分辨率，最佳质量，超细节，专业色彩分级，8K，精确的 ISO 200 设置，焦距 35mm，光圈值 f / 8.0，出图比例 3∶2，版本 v 6.0

其他模板：

模板①提示词：A girl is running, long wavy hair, side profile, hurried, city streets at night, in Hong Kong, face shot, mixed light, neon lights, slow shutter, motion blur, last century style, classic retro, 8K, precise ISO 150 settings, shot on 70mm, f / 8.0, --ar 2:3 --v 5.2

一位女孩在奔跑，长长的卷发，侧面轮廓，匆忙，夜晚的城市街道，在香港，面部拍摄，混合光，霓虹灯，慢快门，动态模糊，上世纪风格，经典复古，8K，精确的 ISO 150 设置，焦距 70mm，光圈值 f / 8.0，出图比例 2∶3，版本 v 5.2

模板②提示词：A photo of an elderly woman, holding crutches on the streets of Hong Kong at night, surrounded by people in motion blur walking past her, motion blur, realistic photography, precise ISO 250 settings, shot on 35mm, f / 11, --ar 3:2 --v 6.0

晚上一位老妇人拄着拐杖走在香港的街道上，周围是一群从她身边走过的模糊人群，动态模糊，真实摄影，精确的 ISO 250 设置，焦距 35mm，光圈值 f / 11，出图比例 3∶2，版本 v 6.0

例：替换成模板①，如图 11.7-2 所示。

图 11.7-2

> **Tips**
>
> 想要拍出动态模糊效果，通常有 3 种方法：①调整曝光时间，我们通常通过长曝光来拍摄移动的物体；②移动相机，制造模糊感，我们通常用于拍摄相对静止的物体；③同时移动相机和物体。

11.8 移轴摄影效果

移轴摄影效果图如图 11.8-1 所示。

图 11.8-1

主题：移轴摄影效果	软件：Midjourney	
光圈值：f / 8.0	感光度：ISO 200	焦距：35mm
色彩：暖色为主	光影：自然光	构图：全景

提示词：Tilt shift photo of Estonia, spring time, in the style of tilt-shift, low light, selective focus, asymmetrical composition, scenery shot, Mediterranean architecture, realistic detail, 8K, precise ISO 200 settings, shot on 35mm, f / 8.0, --ar 16:9 --v 5.2

爱沙尼亚的移轴摄影照片，春天，移轴摄影的风格，低光，选择性聚焦，不对称构图，风景拍摄，地中海建筑，逼真的细节，8K，精确的 ISO 200 设置，焦距 35mm，光圈值 f / 8.0，出图比例 16∶9，版本 v 5.2

其他模板：

模板①提示词：A beautiful view of white buildings, Santorini, blue sea, tilt-shift photography, selective focus, scenery shot, Mediterranean architecture, 8K, precise ISO 150 settings, shot on 35mm, f / 8.0, --ar 16:9 --v 5.2

美丽的白色建筑，圣托里尼，蓝色的大海，移轴摄影，选择性聚焦，风景拍摄，地中海建筑，8K，精确的 ISO 150 设置，焦距 35mm，光圈值 f / 8.0，出图比例 16∶9，版本 v 5.2

模板②提示词：Black and white photograph of the old yellow tram in Budapest, tilt-shift photography, on street, bird's eye view, hyper realistic, precise ISO 250 settings, shot on 35mm, f / 11, --ar 3:2 --v 6.0

布达佩斯旧黄色有轨电车的黑白照片，移轴摄影，在街道上，鸟瞰，超逼真，精确的 ISO 250 设置，焦距 35mm，光圈值 f / 11，出图比例 3∶2，版本 v 6.0

例：替换成模板①，如图 11.8-2 所示。

图 11.8-2

Tips

移轴摄影有很多用途，主要包括控制虚化和控制畸变。例如，在仰拍高楼时，如果出现下大上小的情况，就可以通过移轴来恢复。

11.9 轮廓光效果

轮廓光效果图如图 11.9-1 所示。

图 11.9-1

主题：轮廓光效果　软件：Midjourney

光圈值：f / 5.6　感光度：ISO 200　焦距：70mm

色彩：冷色为主　光影：灯光　构图：近景

提示词：A silhouette of an attractive woman do the ballet, dark room, only her face was illuminated from above by a bright light, black background, contour light, soft shadows, close up portrait, long hair, high contrast, 8K, precise ISO 200 settings, shot on 70mm, f / 5.6, --ar 2:3 --v 6.0

一张在跳芭蕾舞的迷人女人的剪影，黑暗的房间，只有她的脸被明亮的光从上面照亮，黑色的背景，轮廓光，柔和的阴影，近距离肖像，长发，高对比度，8K，精确的 ISO 200 设置，焦距 70mm，光圈值 f / 5.6，出图比例 2∶3，版本 v 6.0

其他模板：

模板①提示词：A woman is silhouetted against the dark, gentle expressions, delicate features, contour light, silhouette lighting, side view, symmetrical composition, monochromatic masterpieces, strong contrast between light and dark, feminine portraiture, extreme details, 8K, precise ISO 150 settings, shot on 80mm, f / 5.6, --ar 3:4 --v 5.2

一个女人的轮廓映衬在黑暗中，温柔的表情，精致的特征，轮廓光，轮廓照明，侧面视图，对称构图，单色杰作，强烈的明暗对比，女性肖像，极致的细节，8K，精确的 ISO 150 设置，焦距 80mm，光圈值 f / 5.6，出图比例 3∶4，版本 v 5.2

模板②提示词：A photo of an attractive woman playing piano in a low light, soft shadows, close up portrait, long hair, contour light, silhouette lighting, precise ISO 250 settings, shot on 70mm, f / 11, --ar 2:3 --v 6.0

一张女人在昏暗的房间里弹钢琴的剪影，柔和的阴影，特写肖像，长发，轮廓光，轮廓照明，精确的 ISO 250 设置，焦距 70mm，光圈值 f / 11，出图比例 2∶3，版本 v 6.0

例：替换成模板①，如图 11.9-2 所示。

图 11.9-2

Tips

想要拍出轮廓光的效果，首先要调整相机的方位，运用正逆光或侧逆光效果的光线，并让主体挡住或接近光源的位置，才能达到刚好让光线勾勒出被拍摄主体的轮廓线的目的。选择不同角度与硬度的光线最终的呈现效果都不尽相同，可以根据拍摄主题多加尝试。

11.10 飞溅效果

飞溅效果图如图 11.10-1 所示。

图 11.10-1

主题：飞溅效果　**软件：**Midjourney

光圈值：f / 8.0　**感光度：**ISO 150　**焦距：**70mm

色彩：冷色为主　**光影：**自然光　**构图：**中景

提示词：A photo of a glass mug with water and lemon slices, surrounded in the style of flying tomatoes, splash moment, dark background, food photography, minimalism, creative photography style, collide with water, 16K, precise ISO 150 settings, shot on 70mm, f / 8.0, --ar 2:3 --v 6.0

一张用玻璃杯盛满水和柠檬片的照片，四周环绕着飞舞的番茄的风格，飞溅的瞬间，深色背景，美食摄影，极简主义，创意摄影风格，与水碰撞，16K，精确的 ISO 150 设置，焦距 70mm，光圈值 f / 8.0，出图比例 2 : 3，版本 v 6.0

其他模板：

模板①提示词：The moment a cherry and blueberry collide with water, with water splashing, bright color, splash moment, black background, layered and complex compositions, commercial style, vivid colors, captured essence of the moment, dynamic and action-packed, 4K, precise ISO 150 settings, shot on 70mm, f / 8.0, --ar 3:2 --v 6.0

樱桃和蓝莓与水碰撞的瞬间，水花飞溅，明亮的色彩，飞溅的瞬间，黑色背景，层次复杂的构图，商业风格，鲜艳的色彩，捕捉瞬间的精华，动感十足，4K，精确的ISO 150设置，焦距70mm，光圈值f / 8.0，出图比例3：2，版本v 6.0

模板②提示词：A can of soda with red liquid splashing around it, splash moment, energetic and vibrant visual effect, high-definition photo, captures every detail of water droplets, dynamic elements, making the drink appear fresh and refreshing, precise ISO 200 settings, shot on 50mm, f / 8.0, --ar 3:4 --v 6.0

一罐溅满红色液体的苏打水，飞溅的瞬间，充满活力的视觉效果，高清照片，捕捉到了水滴的每个细节，动态元素，使饮料显得清新爽口，精确的ISO 200设置，焦距50mm，光圈值f / 8.0，出图比例3：4，版本v 6.0

例：替换成模板①，如图11.10-2所示。

图11.10-2

Tips

为了拍摄飞溅效果，最好将相机设置为连拍模式。为了使拍摄的水滴变模糊，相机需要保持静止，而这需要较短的快门速度，如1/2000等。除此之外，还需要强大的光源或较大的光敏性。为了避免在水面上产生眩光，最好将光源对准背景，而非水滴本身。

第12章　不同构图：超实用的黄金分割构图法

在摄影的过程中，构图是十分重要的环节，对画面效果起着重要的意义。通过规划画面，可以传递出想要表达的特定信息。舍弃画面中普通的、烦琐的、次要的东西，并恰当地安排陪体，可以让画面主体更集中、更典型，以增强艺术效果摄影构图。

12.1 对角线构图

对角线构图效果如图 12.1–1 所示。

图 12.1–1

主题：对角线构图　软件：Midjourney

光圈值：f / 11　感光度：ISO 100　焦距：50mm

色彩：暖色为主　光影：环境光　构图：全景

提示词：In autumn, the river bank is lined with golden ginkgo trees, a child sits at the shore playing happily, the background features a yellow tree in full bloom, creating an atmosphere of warm colors, with high resolution photography, 32K, precise ISO 100 settings, shot on 50mm, f / 11, --ar 2:3 --v 6.0

秋天，河岸两旁种满了金色的银杏树，一个孩子坐在岸边快乐地玩耍，背景是一棵黄色、茂盛的树，营造出暖色调的氛围，高分辨率摄影，32K，精确的 ISO 100 设置，焦距 50mm，光圈值 f / 11，出图比例 2∶3，版本 v 6.0

其他模板：

模板①提示词：Diagonal composition, a photo of an Asian girl sitting at a table, wearing a white long-sleeved shirt and black jeans, she has straight brown hair, there are flowers in vases next to her, 8K, precise ISO 150 settings, shot on 50mm, f / 11,--ar 2:3 --v 6.0

对角线构图，一个亚洲女孩坐在桌子旁，穿着白色长袖衬衫和黑色牛仔裤，有一头棕色的直发，旁边的花瓶里有鲜花，8K，精确的 ISO 150 设置，焦距 50mm，光圈值 f / 11，出图比例 2∶3，版本 v 6.0

模板②提示词：A girl poked her head out from behind the tree, tilted her head, a green tree trunk, in a white dress, leaning against the palm, summer garden background, precise ISO 100 settings, shot on 35mm, f / 2.8, --ar 3:4 --v 6.0

一个女孩从树后探出头来，歪着头，绿色的树干，穿着白色的连衣裙，倚在棕榈树上，夏日花园的背景，精确的 ISO 100 设置，焦距 35mm，光圈值 f / 2.8，出图比例 3∶4，版本 v 6.0

例：替换成模板①，如图 12.1-2 所示。

图 12.1-2

Tips

在运用对角线法则进行构图时，将主体安排在对角线上，会显得主次分明、有立体感，同时给整幅画面增添了运动感和方向感。通常情况下，对角线越长，越能为画面增添延伸感和饱满感。但需要注意的是，画面中的背景部分要尽量干净，否则效果会适得其反。

12.2 框架式构图

框架式构图效果如图 12.2-1 所示。

图 12.2-1

主题：框架式构图　**软件**：Midjourney

光圈值：f / 8.0　**感光度**：ISO 150　**焦距**：50mm

色彩：暖色　**光影**：自然光　**构图**：全景

提示词：Photography, two girls sitting on a bench in the park, looking at white flowers blooming in a garden from inside a black framed arched window with a rounded top, one girl wearing a beige hat is holding another girl’s hand, both carrying backpacks, a pond flowing around them, natural lighting, cinematic composition, 16K, precise ISO 150 settings, shot on 50mm, f / 8.0, --ar 2:3 --v 6.0

摄影，两个女孩坐在公园的长椅上，从一扇圆顶的黑框拱形窗户里看着花园里盛开的白花，一个戴着米色帽子的女孩牵着另一个女孩的手，都背着背包，周围是一个池塘，自然光线，电影构图，16K，精确的 ISO 150 设置，焦距 50mm，光圈值 f / 8.0，出图比例 2：3，版本 v 6.0

其他模板：

模板①提示词：A photo of Mount Fuji taken from the window on an old train in Japan, the view is of houses and city buildings, with the snow capped volcano faintly visible behind them, the back of the child lying on the window, bright colors, sunshine, 32K, precise ISO 200 settings, shot on 50mm, f / 11, --ar 3:2 --v 6.0

一张在日本一列旧火车上从窗口拍摄的富士山照片，照片中可以看到房屋和城市建筑，后面隐约可见白雪覆盖的火山，孩子趴在窗户上，色彩鲜艳，阳光明媚，32K，精确的 ISO 200 设置，焦距 50mm，光圈值 f / 11，出图比例 3：2，版本 v 6.0

模板②提示词：In the rearview mirror of an electric car, people and buildings reflected, the camera captures the scene from above with a wide angle lens, capturing details on both sides of the composition, 16K, precise ISO 150 settings, shot on 50mm, f / 11, --ar 2:3 --v 6.0

从电动汽车的后视镜里，反射出的人和建筑物，相机用广角镜头从上方捕捉场景，捕捉构图两侧的细节，16K，精确的 ISO 150 设置，焦距 50mm，光圈值 f / 11，出图比例 2：3，版本 v 6.0

例：替换成模板①，如图 12.2-2 所示。

图 12.2-2

Tips

框架式构图就是利用各种能形成框架的事物，来作为照片的前景或背景，并区分照片中的主体。在日常生活中，有非常多的场景能够用于框架式构图的创作。框架式构图可以为画面制造纵深感，让照片更加立体直观，更具有视觉冲击力，也可以让主体与环境相呼应。

12.3 引导线构图

引导线构图效果如图 12.3-1 所示。

图 12.3-1

主题：引导线构图　软件：Midjourney
光圈值：f / 8.0　感光度：ISO 150　焦距：35mm
色彩：黑白色　光影：自然光　构图：远景

提示词：Minimalist photography, black and white, woman standing in front of an architectural building wall with geometric shadows cast on it, in the style of Sebastião Salgado, 16K, precise ISO 150 settings, shot on 35mm, f / 8.0, --ar 3:2 --v 6.0

极简主义摄影，黑白，女人站在建筑墙前、墙上投下几何阴影，塞巴斯蒂昂·萨尔加多的风格，16K，精确的 ISO 150 设置，焦距 35mm，光圈值 f / 8.0，出图比例 3∶2，版本 v 6.0

其他模板：

模板①提示词：A car driving on the redwood road in California, with tall trees and yellow lines along both sides of the asphalt highway, taken from behind at eye level, background is a dense green forest, with towering redwood trees lining each side of the two-lane roadway, 8K, precise ISO 200 settings, shot on 35mm, f / 11, --ar 2:3 --v 6.0

一辆汽车行驶在加利福尼亚州的红杉路上，柏油路两边都是高大的树木和黄线，从视线水平的背后拍摄，背景是茂密的绿色森林，高耸的红杉树排列在双车道的两侧，8K，精确的 ISO 200 设置，焦距 35mm，光圈值 f / 11，出图比例 2∶3，版本 v 6.0

模板②提示词：Wide shot, two young Asian children playing on top of a grassy hill, with a clear blue sky, in a cinematic style, with a shutter speed, 16K, precise ISO 150 settings, shot on 35mm, f / 11, --ar 3:2 --v 6.0

广角镜头，两个年轻的亚洲孩子在长满草的山顶上玩耍，天空湛蓝，电影风格，快门速度，16K，精确的 ISO 150 设置，焦距 35mm，光圈值 f / 11，出图比例 3∶2，版本 v 6.0

例：替换成模板①，如图 12.3-2 所示。

图 12.3-2

Tips

引导线构图是吸引视线的有效方法之一，在实际的拍摄中，我们经常会发现一些规律的线条，如河流、车流、光线投影、长廊、街道等。不仅是直线、对角线、弧线，只要是有方向性的、连续的点或线都可以起到引导视觉的作用，都可以称为引导线。利用它们串联画面主体与背景元素，可以更好地引导观众将注意力转移到主体上，并产生深度和透视感。

12.4 对称构图

对称构图效果如图 12.4-1 所示。

图 12.4-1

主题：对称构图　　软件：Midjourney
光圈值：f / 11　　感光度：ISO 150　　焦距：35mm
色彩：暖色为主　　光影：自然光　　构图：全景

提示词：Symmetrical composition, a lady is running on wooden bridges and boardwalks in the endless reeds, high-definition aerial photography, the background features golden yellow grasslands, creating a cinematic feel with natural light and soft tones, 8K, precise ISO 150 settings, shot on 35mm, f / 11, --ar 2:3 --v 6.0

对称构图，一位女士在一望无际的芦苇丛中的木桥和木板路上奔跑，高清航拍，背景以金黄色的草原为特色，用自然光与柔和的色调营造出电影般的感觉，8K，精确的 ISO 150 设置，焦距 35mm，光圈值 f / 11，出图比例 2∶3，版本 v 6.0

其他模板：

模板①提示词：Symmetrical composition, the red and yellow two-colored road in the park, lined with trees and green grass on both sides, leaves spread out after autumn rain, under sunlight, an asphalt track for walking or cycling with white lines painted along it, 4K, precise ISO 100 settings, shot on 50mm, f / 11, --ar 2:3 --v 6.0

对称构图，公园里的红黄两色路旁是树和绿草，秋雨过后落叶铺开，在阳光下，一条用来步行或骑自行车的柏油路上画着白线，4K，精确的 ISO 100 设置，焦距 50mm，光圈值 f / 11，出图比例 2∶3，版本 v 6.0

模板②提示词：A corner of the building with white metal cladding, close up, blue sky, modern minimalist style, high definition photography, symmetrical composition, 16K, precise ISO 200 settings, shot on 100mm, f / 8.0, --ar 3:4 --v 6.0

有着白色金属覆层的建筑一角，近景，蓝天，现代极简风格，高清摄影，对称构图，16K，精确的 ISO 200 设置，焦距 100mm，光圈值 f / 8.0，出图比例 3∶4，版本 v 6.0

例：替换成模板①，如图 12.4-2 所示。

图 12.4-2

Tips

对称构图是较常见的一种构图方式，可以在画面中找到一条明显的中位线将画面分为基本一致的两部分。这种构图方式常用于大多数建筑摄影中，在营造平静、舒适之感的同时，也能体现出严谨、庄重的感觉。

12.5 倒影构图

倒影构图效果如图 12.5–1 所示。

图 12.5–1

主题：倒影构图	软件：Midjourney	
光圈值：f / 8.0	感光度：ISO 150	焦距：35mm
色彩：暖色为主	光影：自然光	构图：远景

提示词：A photo of colorful half–timbered houses along the canals, reflecting on clear water under a blue sky, reflection in water, buildings have detailed wooden beams and pastel colors, enchanting atmosphere, 8K, precise ISO 150 settings, shot on 35mm, f / 8.0, ––ar 3:2 ––v 6.0

运河边五颜六色的半木结构房屋的照片，映照在蓝天下清澈的水面上，水中的倒影，建筑采用细致的木梁和柔和的色彩，迷人的氛围，8K，精确的 ISO 150 设置，焦距 35mm，光圈值 f / 8.0，出图比例 3：2，版本 v 6.0

其他模板：

模板①提示词：A person running on the ground, with reflection of clouds in the water, during sunset, with an orange sky and orange and white clouds, during the golden hour, at mirror lake, in minimalist style, 4K, precise ISO 200 settings, shot on 35mm, f / 11, ––ar 3:2 ––v 6.0

一个人在地面上奔跑，水面上有云的倒影，日落时，橙色的天空和橙白相间的云，黄金时刻，镜湖，极简风格，4K，精确的 ISO 200 设置，焦距 35mm，光圈值 f / 11，出图比例 3：2，版本 v 6.0

模板②提示词：A boy on a bike riding across the beach near clouds, sky–blue and white, reflections, 16K, precise ISO 150 settings, shot on 50mm, f / 8.0, ––ar 3:2 ––v 6.0

一个男孩骑着自行车穿过海滩，附近有云，天蓝色和白色，倒影，16K，精确的 ISO 150 设置，焦距 50mm，光圈值 f / 8.0，出图比例 3：2，版本 v 6.0

例：替换成模板①，如图 12.5-2 所示。

图 12.5-2

Tips

倒影是摄影的对称构图中应用较多的方式之一。除常见的水面外，镜面、雨后潮湿的路面、冰面等，都可以制造出绝佳的倒影效果。在拍摄的过程中，我们应遵守化少为多、化繁为简的基本规则，保留最重要的元素，即主体和倒影，而将其余的干扰元素移出画面。这样才能让拍摄画面简单而内涵丰富。

除此之外，掌握拍摄角度同样很关键，较远处的倒影与实景比例几乎是等大的，越靠近的倒影，则因视点高度的不同，显示其多少之分，倒影的多少与视点高度成反比。总而言之，拍摄视角高，倒影显得少；拍摄视角低，倒影显得多。

12.6 曲线构图

曲线构图效果如图 12.6-1 所示。

图 12.6-1

主题：曲线构图　　软件：Midjourney
光圈值：f / 11　　感光度：ISO 100　　焦距：35mm
色彩：冷色和暖色相结合　　光影：自然光　　构图：全景

提示词：Aerial photo, a woman in a red dress and high heels, sit on a spiral staircase, with black hair, in an elegant pose, on a white marble floor, white walls, in a minimalist style, with natural light, 8K, precise ISO 100 settings, shot on 35mm, f / 11, --ar 2:3 --v 6.0

航拍照片，一个身穿红色长裙和高跟鞋的女人，坐在螺旋楼梯上，一头黑发，摆出优雅的姿势，在白色大理石地板上，白色墙壁，极简风格，自然光，8K，精确的 ISO 100 设置，焦距 35mm，光圈值 f / 11，出图比例 2∶3，版本 v 6.0

其他模板：

模板①提示词：The advanced lighting of the bridge in the far view, with a wide angle lens, a city background at night, bright lights, modern architecture and spectacular scenery, bright light illuminates bridge structures and urban landscapes, architectural grandeur, 4K, precise ISO 200 settings, shot on 35mm, f / 5.6, --ar 3:2 --v 6.0

远处桥梁的照明，广角镜头，夜晚的城市背景，明亮的灯光，令人印象深刻的现代建筑和壮观的景色，明亮的灯光照亮了桥梁结构和城市景观，建筑的宏伟，4K，精确的 ISO 200 设置，焦距 35mm，光圈值 f / 5.6，出图比例 3∶2，版本 v 6.0

模板②提示词：A photo of winding roads in front of snow capped mountains and lakes, clean background, minimal style, cool color tones, high resolution photography, natural lighting and professionally color graded, with soft shadows and smooth details, 8K, precise ISO 150 settings, shot on 35mm, f / 8.0, --ar 3:4 --v 6.0

一张在雪山和湖泊前蜿蜒的道路照片，干净的背景，极简的风格，冷色调，高分辨率的摄影，自然照明和专业的色彩渐变，柔和的阴影和光滑的细节，8K，精确的 ISO 150 设置，焦距 35mm，光圈值 f / 8.0，出图比例 3∶4，版本 v 6.0

例：替换成模板①，如图 12.6-2 所示。

图 12.6-2

Tips

曲线构图是指将画面中的重要元素沿曲线排布，其他元素填充剩余空间。曲线具有灵活的属性，可以延长并变化画面，从前景向中景和后景延伸，画面构成纵深方向的空间关系的视觉感，给画面制造韵律感，产生优美、雅致、协调的感觉。

12.7 三分法构图

三分法构图效果如图 12.7–1 所示。

图 12.7–1

主题：三分法构图　**软件：**Midjourney
光圈值：f / 11　**感光度：**ISO 150　**焦距：**35mm
色彩：冷色为主　**光影：**自然光　**构图：**全景

提示词：A girl is painting on the beach, holding an easel, rule of thirds, with rocks at sea level and a blue sky as background, high definition and clear, 8K, precise ISO 150 settings, shot on 35mm, f / 11, --ar 3:4 --v 6.0

一个女孩正在沙滩上画画，手里拿着画架，三分法，以海平面上的岩石和蓝天为背景，高清晰度和清晰的，8K，精确的 ISO 150 设置，焦距 35mm，光圈值 f / 11，出图比例 3∶4，版本 v 6.0

其他模板：

模板①提示词：A girl is reading in a bookstore, holding books and smiling, the bookshelf behind her has various types of picture storybooks, she stands between two shelves, captured from a frontal perspective, rule of thirds, high resolution and sharp focus, 4K, precise ISO 200 settings, shot on 50mm, f / 11, --ar 2:3 --v 6.0

一个女孩在书店里看书，手里拿着书并微笑着，身后的书架上有各种各样的绘本故事书，她站在两个架子之间，从正面角度拍摄，三分法，高分辨率和清晰的焦点，4K，精确的 ISO 200 设置，焦距 50mm，光圈值 f / 11，出图比例 2：3，版本 v 6.0

模板②提示词：In the evening, there is sand and grass around it, a young Chinese boy stands by the river with his back to us, wearing blue shorts and white long sleeves, a sense of atmosphere in the style of cinematography, rule of thirds, 8K, precise ISO 150 settings, shot on 35mm, f / 8.0, --ar 3:4 --v 6.0

晚上，周围是沙子和草，一个年轻的中国男孩背对着站在河边，穿着蓝色短裤和白色长袖上衣，电影风格的氛围感，三分法，8K，精确的 ISO 150 设置，焦距 35mm，光圈值 f / 8.0，出图比例 3：4，版本 v 6.0

例：替换成模板①，如图 12.7-2 所示。

图 12.7-2

Tips

三分法是一种偏离中心的构图方式，将画面中的重要元素沿着 3×3 的网格排列。这种构图方式可以赋予照片结构并使其更具吸引力。三分法是一条法则，但不是一条规则，不需要生硬地搬到每一个场景中。在理解原理后根据具体情况应用，才能拍出好的作品。

12.8 留白构图

留白构图效果如图 12.8-1 所示。

图 12.8-1

主题：留白构图	软件：Midjourney	
光圈值：f / 11	感光度：ISO 150	焦距：50mm
色彩：冷色	光影：自然光	构图：全景

提示词：A small boat in the middle of calm waters, the silhouette of a person rowing against the cliff wall, minimalist style, black and white photography, water reflection, wide-angle lens, soft light, static scene, peaceful atmosphere, 8K, precise ISO 150 settings, shot on 50mm, f / 11, --ar 2:3 --v 6.0

一条小船在平静的水面中间，一个人在崖壁下划船的剪影，极简风格，黑白摄影，水面反射，广角镜头，柔和的光线，静态的场景，宁静的气氛，8K，精确的 ISO 150 设置，焦距 50mm，光圈值 f / 11，出图比例 2∶3，版本 v 6.0

其他模板：

模板①提示词：Wide shot, a woman standing on a grassy hill, wearing a hat and white dress, clear blue sky, sunlight, 4K, precise ISO 200 settings, shot on 35mm, f / 11, --ar 3:2 --v 6.0

广角镜头，一个女人站在长满草的小山上，戴着帽子并穿着白色裙子，湛蓝的天空，阳光，4K，精确的 ISO 200 设置，焦距 35mm，光圈值 f / 11，出图比例 3∶2，版本 v 6.0

模板②提示词：In the foreground is an endless expanse of water, with clear blue sky and light clouds in the background, a man stands on one side of it, minimalist composition, photography, 8K, precise ISO 200 settings, shot on 35mm, f / 8.0, --ar 4:5 --v 6.0

前景是一望无际的汪洋大海，背景是湛蓝的天空和淡淡的云彩，一个身穿黑衣的男人站在海的一边，极简主义的构图，摄影，8K，精确的 ISO 200 设置，焦距 35mm，光圈值 f / 8.0，出图比例 4∶5，版本 v 6.0

例：替换成模板①，如图 12.8-2 所示。

图 12.8-2

Tips

留白的内容不一定是白色的画面，还可能是天空、水面、雾色、近颜色的物体等。在画面中增加适量的留白可以制造元素多与少、疏与密的对比，从而增强节奏韵律感。此外，这样的开放式构图，还能引人遐想。

12.9 中心构图

中心构图效果如图 12.9-1 所示。

图 12.9-1

主题：中心构图	软件：Midjourney	
光圈值：f / 11	感光度：ISO 150	焦距：35mm
色彩：暖色和冷色相结合	光影：自然光	构图：全景

提示词：A small boat is rowing on the river, with white cherry blossoms blooming beside it, green trees in spring by its side, there's an ancient house in the style of Chinese style next to the water, photography style, high definition, wide-angle lens, bright colors, calm environment, people sitting inside boats enjoying the scenery, 8K, precise ISO 150 settings, shot on 35mm, f / 11, --ar 2:3 --v 6.0

一条小船在江面上，一边是盛开的白色樱花，一边是春天的绿树，水边有一座中式风格的古宅，摄影风格，高清，广角镜头，色彩明快，环境平静，人们坐在船里欣赏风景，8K，精确的 ISO 150 设置，焦距 35mm，光圈值 f / 11，出图比例 2∶3，版本 v 6.0

其他模板：

模板①提示词：Centered composition, a photo of modern ships and fishing boats at sea, with seagulls flying overhead, shore with lots of blue water, during the golden hour, cinematic feel, with vibrant colors, high resolution, in a hyper realistic style, 4K, precise ISO 200 settings, shot on 35mm, f / 11, --ar 2:3 --v 6.0

中心构图，一张现代船只和渔船在海上的照片，海鸥从头顶飞过，有大量蓝色海水的海岸，黄金时刻，电影般的感觉，色彩鲜艳，高分辨率，超现实主义的风格，4K，精确的ISO 200 设置，焦距 35mm，光圈值 f / 11，出图比例 2：3，版本 v 6.0

模板②提示词：A fisherman in small boat is fishing at sea during sunset, right in the middle, the sky has an orange and red color, reflecting on the calm water surface, the vast ocean stretches to the horizon, creating a serene scene of solitude, ultra realistic photo, cinematic style, 8K, precise ISO 200 settings, shot on 50mm, f / 8.0, --ar 4:3 --v 6.0

黄昏时一个渔夫在海面的小船上捕鱼，正中央，天空呈现出橙红色，倒映在平静的水面上，辽阔的海洋一直延伸到地平线，创造了一个宁静而孤独的场景，超现实主义的照片，电影风格，8K，精确的 ISO 200 设置，焦距 50mm，光圈值 f / 8.0，出图比例 4：3，版本 v 6.0

例：替换成模板①，如图 12.9-2 所示。

图 12.9-2

Tips

中心构图法可以将最核心的内容直观地展示给观众，重点突出、主次分明，很好地提高了信息的传达效率，并引导观众的视线聚集在想要突出的内容上。在不知道如何构图时，这是一种比较万能的构图法则。

后期制作篇

第13章　换　　脸

AI 换脸是指利用 AI 技术，提取人脸特征，从而将一张人脸迁移到另一张人脸上，实现人脸转换的过程。目前这项技术已被广泛应用于各个领域中，如视频制作、艺术修复等。如果我们也想要尝试 AI 换脸，将自己的照片应用于不同风格中，不少软件可以辅助我们进行操作。

13.1 常用的换脸工具介绍

目前，常用的几种换脸工具如下。

- 智能修复老照片

如图 13.1–1 所示，智能修复老照片是一个照片修复工具，在对褪色、模糊的照片进行修复的基础上，还可以给照片添加很多趣味特效，其中包含“AI 百变形象”，即换脸的功能。其操作也很简单，只需单击“AI 百变形象”功能，选择自己喜欢的风格，并添加自己的照片，就可以自行换脸了。

图 13.1–1

- 美图秀秀

如图 13.1–2 所示，美图秀秀是一款常用的图片处理软件，可以帮助我们对照片进行多项后期处理。其中也包含了“AI 写真”，即换脸的功能。单击“AI 写真”功能，上传一张自己的照片，创建个人的面部档案。随后就可以自行选择喜欢和想要的风格，实现换脸效果。

图 13.1–2

- faceswap

如图 13.1–3 所示，faceswap 是一款 AI 换脸软件。选择自己的照片和想要的图像风格，软件将自动将原图像人脸的特征转移到目标图像上，从而实现换脸效果。除此之外，faceswap 还支持视频换脸功能，在换脸的基础上还能保留原视频的动态效果。

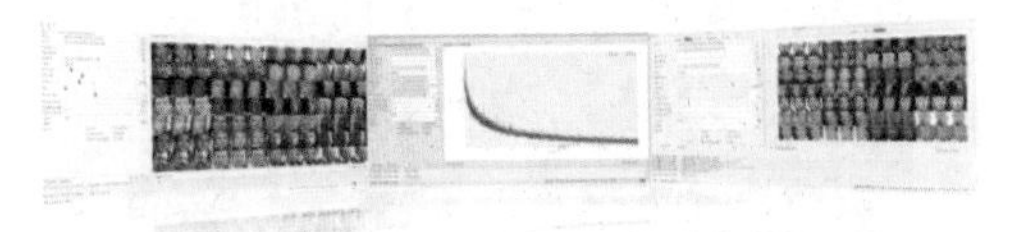

图 13.1–3

- Midjourney

如图 13.1–4 所示，Midjourney 是一款 AI 绘画工具，可以在输入的提示词文本上，通过 AI 算法迅速生成具有绘画性和美观性的图像。其中也包含了 AI 换脸功能，Midjourney 的换脸操作相对复杂，将在 13.2~13.4 节中进行详细讲解。

图 13.1–4

13.2 制服照换脸

如果我们急需一张属于自己的制服照，又临时找不到合适的场所或不方便出去拍，这时，Midjourney 的换脸功能可以为我们提供很大的帮助。那么，具体要怎么操作呢？下面以实践篇 7.7 节中的图像为例介绍以下案例。

13.2.1 最终效果图

最终效果图如图 13.2-1 所示。

图 13.2-1

13.2.2 步骤

步骤 01 如图 13.2-2 和图 13.2-3 所示，在浏览器中打开 InsightFace 的授权链接，并给自己的服务器添加机器人。

图 13.2-2

图 13.2-3

步骤 02 在成功添加 InsightFace 机器人后，就可以上传自己的照片了。如图 13.2-4 所示，单击 Midjourney 对话框，输入“/”后选择“/saveid”功能。

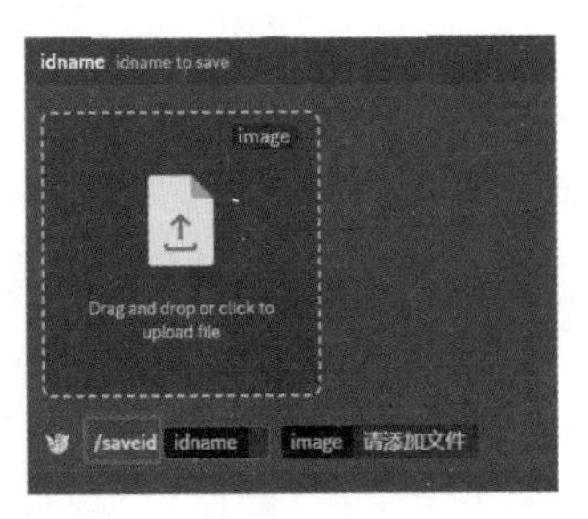

图 13.2-4

步骤 03 如图 13.2-5 和图 13.2-6 所示，将自己的照片拖进虚线框，可以在 idname 栏中为这张照片起个名字用于区分，但注意控制在 10 个字符以内。按 Enter 键发送指令后会显示该 ID 已创建，如图 13.2-7 所示。

图 13.2-5

图 13.2-6

图 13.2-7

步骤 04 如图 13.2-8 和图 13.2-9 所示，上传自己想要的制服照片。右击图像，在弹出的快捷菜单中选择 APP → INSwapper 命令，即可生成属于自己的制服照。单击大图，选择用浏览器打开，在网页内右击保存图像。完成后的效果见图 13.2-1。

图 13.2-8

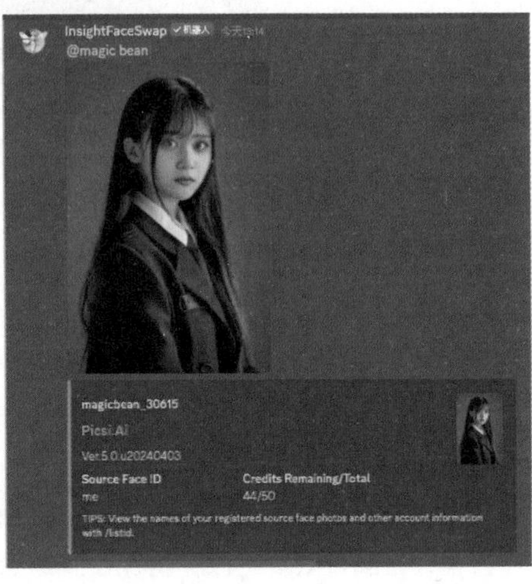

图 13.2-9

13.3 艺术照换脸 1

如果想要一张属于自己的艺术照，又找不到合适的风格，想要多加尝试，这时，Midjourney 的换脸功能可以提供很大的帮助。那么，具体要怎么操作呢？下面以实践篇 7.8 节中的图像为例介绍以下案例。

13.3.1 最终效果图

最终效果图如图 13.3–1 所示。

图 13.3–1

13.3.2 步骤

步骤 01 单击 Midjourney 对话框，输入“/”后选择“/saveid”功能。

步骤 02 将自己的照片拖进虚线框，可以在 idname 栏中为这张照片起个名字用于区分，但注意控制在 10 个字符以内。按 Enter 键发送指令后会显示该 ID 已创建。

步骤 03 如图 13.3–2 和图 13.3–3 所示，上传自己想要的艺术照。右击图像，在弹出的快捷菜单中选择 APP → INSwapper 命令，即可生成属于自己的艺术照。单击大图，选择用浏览器打开，在网页内右击保存图像。完成后的效果见图 13.3–1 所示。

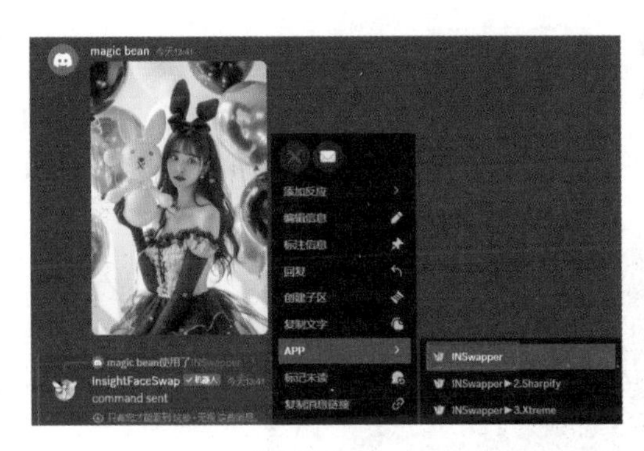

图 13.3-2

图 13.3-3

13.4 艺术照换脸 2

如果想要一张属于自己的日式小清新风格的照片，身边又没有合适的服装且找不到合适的场地拍摄，这时，Midjourney 的换脸功能可以提供很大的帮助。那么，具体要怎么操作呢？下面以实践篇 9.9 节中的图像为例介绍以下案例。

13.4.1 最终效果图

最终效果图如图 13.4-1 所示。

图 13.4-1

13.4.2 步骤

步骤 01 单击 Midjourney 对话框，输入“/”后选择“/saveid”功能。

步骤 02 将自己的照片拖进虚线框，可以在 idname 栏中为这张照片起个名字用于区分，但注意控制在 10 个字符以内。按 Enter 键发送指令后会显示该 ID 已创建。

步骤 03 如图 13.4–2 和图 13.4–3 所示，上传自己想要的风格照。右击图像，在弹出的快捷菜单中选择 APP → INSwapper 命令，即可生成属于自己的日式小清新风格照。单击大图，选择用浏览器打开，在网页内右击保存图像。完成后的效果见图 13.4–1。

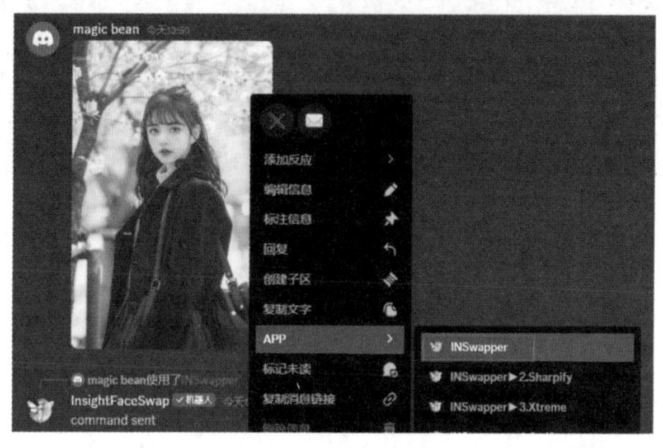

图 13.4–2

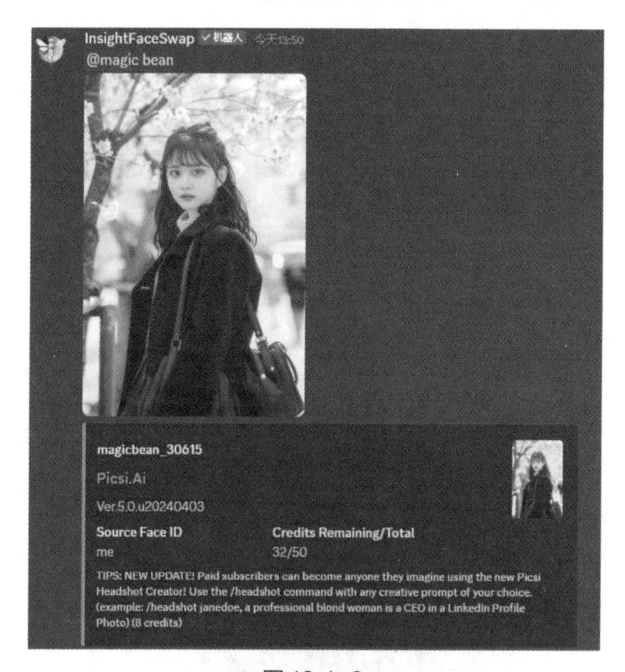

图 13.4–3

第14章　后 期 处 理

后期处理工具在摄影中起到了很大的作用，它能够辅助修饰图像、矫正画面、添加特效等，是摄影师表现创意和思想的重要形式之一。除此之外，后期处理工具还可以用于修复老旧的照片，并已经逐渐成为摄影中一个必不可少的工具和过程。

14.1 常用的后期处理工具介绍

目前，常用的几种后期处理工具如下。

- Photoshop

如图 14.1–1 所示，Photoshop 是一款功能非常强大的图像处理软件，主要用于处理由像素构成的数字图像。用户可以使用其众多的编修与绘图工具，进行选择、合成和调色，从而有效地进行图像编辑和修饰。该工具普遍运用于广告设计、美术创意和三维动画行业。

图 14.1–1

- Lightroom

如图 14.1–2 所示，Lightroom 是一款以后期制作为重点的图像编辑软件。其中包含图像管理、修改编辑、输出等功能，是一套完整的后期处理流程。主要面向数码摄影、图形设计等专业领域，如婚纱摄影等。

图 14.1–2

- 美图秀秀

如图 14.1–3 所示，美图秀秀是一款手机端图像编辑 App，主要用于图像的加工处理。其操作简单、功能齐全，可以用于图像的美化，如滤镜、加贴纸、调整参数等；同时还提供人像的美化，如瑕疵修复、面部重塑等。

图 14.1–3

14.2 用手机自带编辑功能处理

如果想要用手机自带的编辑功能对 Midjourney 生成的图像进行加工，具体要怎么操作呢？下面以实践篇 5.6 节中的图像为例介绍以下案例。

14.2.1　最终效果图

最终效果图如图 14.2-1 所示。

图 14.2-1

14.2.2　步骤

步骤 01　在相片库中打开 Midjourney 生成的图像，单击“编辑”功能。在修剪选项中将图像裁剪调整为需要的尺寸，如图 14.2-2 所示。

图 14.2-2

步骤 02　在“调节”选项中调整图像的饱和度，如图 14.2-3 所示。

步骤 03　在“滤镜”选项中选择想要的滤镜风格，如图 14.2-4 所示。

图 14.2-3

图 14.2-4

步骤 04 单击“对比”按钮，可以直观地看到原图与效果图的区别。编辑好图像后确认，并保存图像即可。完成后的效果见图 14.2-1。

14.3 用美图秀秀处理

如果想要用美图秀秀对 Midjourney 生成的图像进行加工，具体要怎么操作呢？下面以实践篇 7.9 节中的图像为例介绍以下案例。

14.3.1 最终效果图

最终效果图如图 14.3-1 所示。

图 14.3-1

14.3.2 步骤

步骤 01 将 Midjourney 生成的图像置入美图秀秀，选择图像美化功能。观察图像后，可以发现整体色调偏暗。单击“调色”功能，在“光效”栏中根据需求依次调整亮度、对比度、曝光、光感，具体数值如图 14.3-2 所示。在“色彩”栏中根据需求调整色温，具体数值如图 14.3-3 所示。在“细节”栏中根据需求调整暗角，具体数值如图 14.3-4 所示。

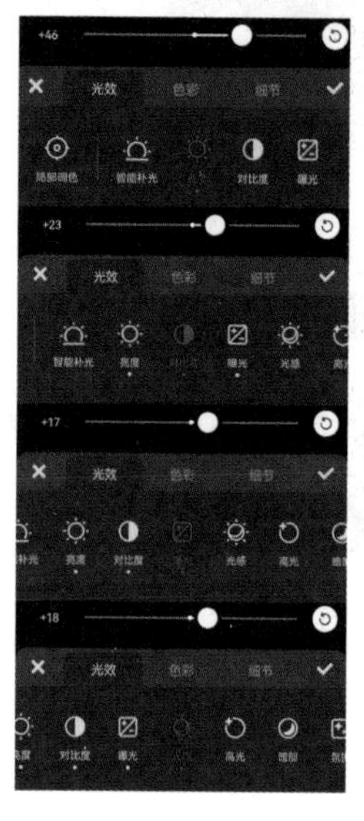

图 14.3-2

图 14.3-3

图 14.3-4

步骤 02 单击“人像美容”功能，选择美妆。根据自己的需求对人物进行妆容调整，具体数值如图 14.3-5 所示。

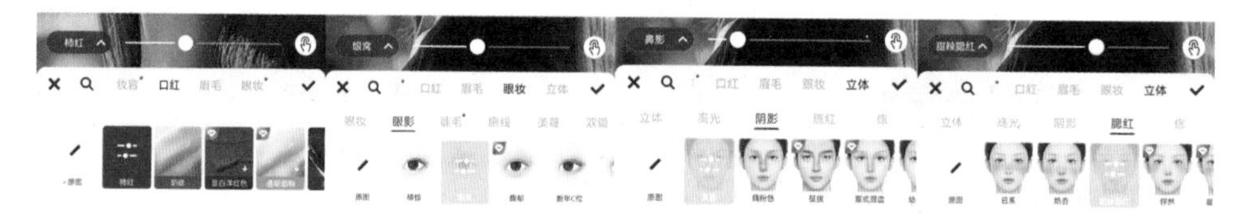

图 14.3-5

步骤 03 单击视图，可以直观地看到原图与效果图的区别。编辑好图像后确认，并保存图像即可。完成后的效果见图 14.3-1。

14.4 用 Photoshop 处理

如果想要用 Photoshop 对 Midjourney 生成的图像进行加工，具体要怎么操作呢？下面以实践篇 3.1 节中的图像为例介绍以下案例。

14.4.1 最终效果图

最终效果图如图 14.4-1 所示。

图14.4–1

14.4.2 步骤

步骤 01 将 Midjourney 生成的图像置入 Photoshop。单击“滤镜”功能，找到 Camera Raw 滤镜，对图像进行处理，如图 14.4–2 所示。

图14.4–2

步骤 02 观察图像，可以发现整体色调偏暗，没有春天的明媚感。如果需要一张偏亮的图像，可以自行调整曝光效果。在调整完成后，还可以增加一些饱和度和晕影，加强画面的冲击力。具体数值如图 14.4–3 和图 14.4–4 所示。

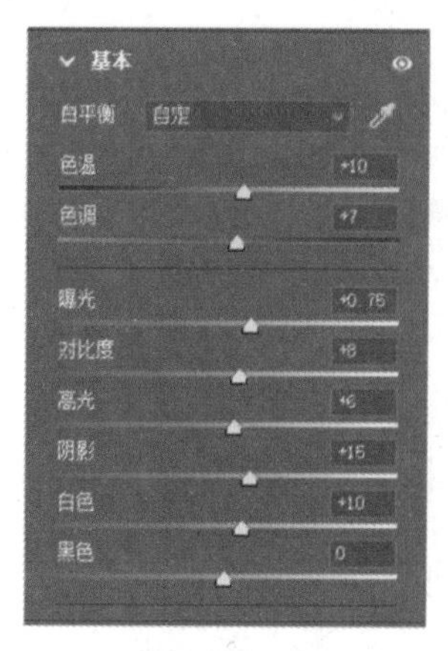

图 14.4-3

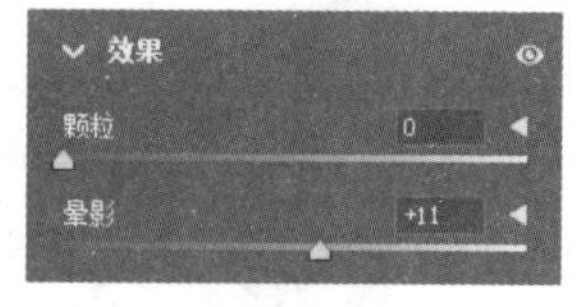

图 14.4-4

步骤 03 单击视图，可以直观地看到原图与效果图的区别，如图 14.4-5 和图 14.4-6 所示。编辑好图像后单击“确认”按钮，并导出图像即可。完成后的效果见图 14.4-1。

图 14.4-5

图 14.4-6